Tamarind
Science and Technology

The Authors

Dr. A. Balasubramanian, Professor and Head, Department of Silviculture, Forest College and Research Institute, Tamil Nadu Agricultural University, Mettupalayam. He holds rich experience in forestry research and has played a primordial role by handling more than 21 research projects. He is guiding for courses on silviculture, forest ecology and biodiversity, forest and climate change at under graduate, post graduate and doctoral level. He has published six books and furnished more than 150 research articles. He has guided over 25 students of M.Sc. (Forestry) and Ph.D. (Forestry) research. Dr. Balasubramanian established modern research facilities for forestry research in his institute including mobile ban based continuous ambient air quality monitoring system (CAAQMS). He has been awarded with more than 15 awards and honors of national and international repute for his achievement in forestry research and education.

Dr. P. Sudha, is working as Assistant Professor (FPE). Her field of specialization is Food and Agricultural Process Engineering. She has publications at national and international level in her research space. She has published various book chapters. Previously she was working at Faculty of Engineering, Avinashilingam University. Her area of research is on food hydro colloids.

Dr. C. N. Hari Prasath, is presently working as Senior Research Fellow, Department of Silviculture, Forest College and Research Institute, Tamil Nadu Agricultural University, Mettupalayam. His field of specialization includes carbon sequestration, precision silviculture, natural gums and commercial tree farming. He has published more than 20 research articles in national and international journals in the field of Forestry Research. He worked as a Junior Research Fellow in Department of Science and Technology on a funded scheme on Afforestation and Carbon Sequestration. Dr. Prasath also worked as Senior Research Fellow in ICAR- Indian Institute of Natural Resins and Gums funded scheme on Harvesting, Processing and Value Addition of Natural Resins and Gums. He is honored with several prestigious awards for his contribution in the field of forestry research.

M. Sangareswari Nagajothi, is a doctoral student from Forest College and Research Institute, Mettupalayam. She is specialises in silviculture and has completed her dissertation work on 'Evolving silvicultural and processing techniques for enhancing tamarind fruit yield and gum production'.

Dr. S. Radhakrishnan, is presently working as Assistant Professor (Forestry), Department of Silviculture, Forest College and Research Institute, Tamil Nadu Agricultural University, Mettupalayam. He holds 12 years of teaching and research experience. He has handled five research projects. He has handled more than 80 courses for students in under graduate, post graduate and doctoral level. He has guided 5 students for their M.Sc. (Forestry) programme. To his credit, he has published five books and more than 100 research papers in international and national journals, international and national conferences, seminars and symposiums. Dr. Radhakrishnan is honored with five awards for his contribution in the field of forestry research and education.

Dr. K. K. Suresh, Professor (Forestry) and Former Dean, Forest College and Research Institute, Tamil Nadu Agricultural University, Mettupalayam. He has 31 years of experience in teaching, research and education in forestry. He is trained in Australia on forest inventory, and his field of specialization includes precision silviculture and commercial tree farming. He was instrumental in developing certification standards for quality planting material in tree species. He has guided 15 post graduate students in forestry and has over 40 research articles published in journals of national and international repute, and four books to his credit. His commendable work in association with the 'National Bamboo Mission' on clonal multiplication and promotion of thorn less bamboo among farmers was greatly applauded across India.

Tamarind
Science and Technology

A. Balasubramanian
P. Sudha
C.N. Hari Prasath
M. Sangareswari
S. Radhakrishnan
K.K. Suresh

SCIENTIFIC
PUBLISHERS

Published by
SCIENTIFIC PUBLISHERS (INDIA)

Jodhpur –

5 A, New Pali Road
P.O. Box 91
Jodhpur - 342 001 INDIA

Delhi –

4806/24, Ansari Road
Daryaganj
New Delhi - 110 002 INDIA

ISBN: 978-93-86652-25-6

Visit the Scientific Publishers (India) website at
http://www.scientificpub.com

Printed in India

Dedicated

to

Tamil Nadu Agricultural University

(TNAU)

&

Indian Council of Agricultural Research

(ICAR)

TAMIL NADU AGRICULTURAL UNIVERSITY

Dr. K. Ramasamy, Ph.D.,
Vice-Chancellor

Coimbatore - 641 003
Tamil Nadu, India

FOREWORD

Tamarind being a multipurpose tree is predominantly found in farmlands, forest area and road avenues. India's annual tamarind production is about 200-210 MT, Tamil Nadu and Karnataka alone contribute over 60 per cent of the country's production. Besides tamarind is commonly grown in the states of Andhra Pradesh, Maharashtra, Telangana, Kerala, Bihar, Orissa, Chhattisgarh, Pondicherry, Mizoram, Andaman & Nicobar.

Tamarind pod contains 40% pump, 26% seed, 3% fiber and 31% pod shell. Tamarind pulp is an important ingredients in food recipes of Southern India. Although, the utility of tamarind pulp is well known, but the seed as source of gum is rarely known. The tamarind seed contains "Jellose" (Gummy polysaccharide) is used as edible gum. With the advent of modern scientific technologies as high as 52% gum is extracted from one kilogram of Tamarind Kernal Powder (TKP).

In order to exploit the benefit of Tamarind seed gum, Forest College and Research Institute, Tamil Nadu Agricultural University, Mettupalayam has joined with ICAR - Indian Institute of Natural Resins and Gums, Ranchi through the Network project on "Harvesting Process and Value Addition of Natural Resins and Gums". Various research findings emanated through the project have helped to publish the book on **"Tamarind Science and Technology"** by a team of scientists from Forest College and Research Institute, Mettupalayam.

The book contains information on biology and silvicultural management of Tamarind, besides it enlightens informations on harvesting, processing and value addition in Tamarind with particular reference to Tamarind seed gum. Thus, a totally information of Tamarind from production to supply chain was brought out in this book.

I strongly believe that the book will be very much useful to researchers, students, industries, farmers and other stakeholders. I congratulate the effort of Dr. A. Balasubramanian, Dr. P. Sudha, Dr. C.N. Hari Prasath, Dr. M. Sangareswari, Dr. S. Radhakrishnan and Dr. K.K. Suresh for their concerted effort to publish this book. I also congratulate to Mr. Tanay Sharma for Scientific Publishers (India), Jodhpur to publish this book in a most readable format. I am thankful to Indian Council of Agricultural Research (ICAR), Indian Institute of Natural Resins and Gums (IINRG), Ranchi for their support in publishing the book.

(K. RAMASAMY)

• Tel: Off.: +91 422 2431788 • Res.: +91 422 2430887 • Fax: +91 422 2431672 •Email: vctnau@tnau.ac.in

Acknowledgement

We whole heartedly express our deep sense of gratitude and indebtness to **Dr. K. Ramasamy**, Vice chancellor, Tamil Nadu Agricultural University, Coimbatore for providing facilities and extending the guidance in publishing this book through the implementation of ICAR Network Project on "Harvesting, Processing and Value Addition of Natural Resins and Gums" at Forest College and Research Institute, Mettupalayam. We record our sincere thanks to Tamil Nadu Agricultural University (TNAU) and Forest College and Research Institute (FC&RI) for providing excellent platform in publishing the book. We equally thankful to Indian Council of Agricultural Research (ICAR) and Indian Institute of Natural Resins and Gums (IINRG), Ranchi for shaping the book. It is rather a pleasant privilege to place our deepest sense of gratitude and affection to Dr. Sharma, Director, IINRG, Ranchi; Dr. Ramani, Former Director, IINRG, Ranchi and Dr. Niranjan Prasad, Principal Scientist & Network Coordinator of HPVA-NRG, IINRG, Ranchi for rendering the ideas by providing information and help in shaping this book.

We acknowledge the contribution by different experts for their valuable scientific information as book chapters compiled in this book. We sincerely thank the Mr. Tanay Sharma for Scientific Publishers (India), Jodhpur for neat execution in publishing this book.

- AUTHORS

List of Contributors

Dr. E. Alli Rani, Ph.D.,
Research Associate,
Department of Vegetable Crops,
HC& RI, Tamil Nadu Agricultural University,
Coimbatore – 641 003

Dr. A. Balasubramanian, Ph.D.,
Professor and Head,
Department of Silviculture,
Forest College and Research Institute,
Mettupalayam – 641 301

Dr. M. Chinnadurai, Ph.D.,
Director,
Centre for Agricultural and Rural Development studies
Tamil Nadu Agricultural University,
Coimbatore – 641 003 .

Dr. P. Chitra, Ph.D.,
Assistant Professor (V& AS)
Department of Silviculture
Forest College and Research Institute
Mettupalayam - 641301

Dr. C.N. Hari Prasath, Ph.D.,
Senior Research Fellow (SRF),
Department of Silviculture,
Forest College and Research Institute,
Mettupalayam – 641 301

Dr. I. Jaisankar, Ph.D.
Senior Scientist
Department of Horticulture and Forestry, Central Island Agriculture Research Institute
Andaman & Nicobar Islands

Dr. N. Karpoora Sundara Pandian, Ph.D.
Assistant Professor,
College of Food and Dairy Technology,
Tamil Nadu Veterinary and Animal Sciences University,
Chennai – 600 052

Dr. V. Ponnuswami Ph.D., PDF (Taiwan)
Former Dean & Professor (Horticulture),
Horticultural College & Research Institute,
Tamil Nadu Agricultural University,
Coimbatore – 641 003

Dr. P. Pretheep Kumar, Ph.D.,
Assistant Professor (Agrl . Entomology)
Department of Forest Products and Utilization, Forest College and Research Institute,
Mettupalayam – 641 301

Dr. S. Radhakrishnan, Ph.D.
Assistant Professor (Forestry),
Department of Silviculture,
Forest College and Research Institute,
Mettupalayam – 641 301

Dr. P. Rajkumar, Ph.D.
Professor and Head,
Department of Agricultural Processing and Basic Sciences,
Agricultural Engineering College and Research Institute,
Kumulur, Trichy – 621 712

Dr. Reni, Ph.D .
Assistant Professor
Department of Food Processing and Preservation Technology,
Avinashilingam University,
Coimbatore – 641 003

Dr. P. Sudha, Ph.D.
Assistant Professor (FPE),
Centre for Excellence in Biofuels
Forest College and Research Institute, Mettupalayam – 641 301

Ms. M. Sangareswari Nagajothi,
Research Scholar,
Department of Silviculture,
Forest College and Research Institute, Mettupalayam – 641 301

Dr. M. Shakila Banu, Ph.D.,
Professor and Head
Department of Food Processing and Preservation Technology
Avinashilingam University
Coimbatore – 641 003

Dr. S. Varadha Raj, Ph.D.,
Assistant Professor
(Agricultural Economics)
Tamil Nadu Agricultural University
Coimbatore – 641 003

Contents

Chapter *1*

TAMARIND – It's Introduction

Dr. C.N. Hari Prasath*, Dr. A. Balasubramanian & Dr. S. Radhakrishnan*****

**Senior Research Fellow, Department of Silviculture, Forest College and Research Institute, Mettupalayam*
***Professor and Head, Department of Silviculture,*
**** Assistant Professor (Forestry), Department of Silviculture*
Forest College and Research Institute, Mettupalayam

Tamarind (*Tamarindus indica* L.) is a multipurpose tropical fruit tree used for its fruits (Eaten fresh or processed), seeds (Food and non-food uses) and wood. Tamarind belongs to the dicotyledonous, family Leguminosae and sub-family Caesalpinioideae. *Tamarindus* is a monotypic genus, containing the sole species *T. indicus* (Fig. 1.1).

Fig. 1.1. Tamarind Plantation

1.1. ORIGIN AND DISTRIBUTION

Tamarind is one of the most common trees of tropical India; which is believed to be indigenous to tropical Africa and probably introduced long back into India by Arabs. The fruit was well known to the ancient Egyptians and to the Greeks in the 4th century B.C. It is not the tree of the forests but is cultivated throughout the country, except the Himalayan region and Arid Western Zone. In Maharashtra, Bihar, Odisha, Karnataka, Tamil Nadu and Andhra Pradesh, it is the principal tree of the village commons and extensively planted as a roadside shade tree. It is essentially that a tree of the tropical climate, does not flourish and produce ripen fruits. It is also cultivated in Bangladesh, Srilanka, Burma and several other parts of the world viz. Florida (U.S.A.), Sudan, Egypt, Taiwan, Malaysia and South East Asian countries. It has also been introduced in Australia and South America (El-Siddig *et al.*, 2006).

The tree has long been naturalized in the East Indies and the islands of the Pacific, which was early introduced into tropical America and the West Indies. In all tropical and near-tropical areas, including South Florida, it is grown as a shade and fruit tree, along roadways, dooryards and parks. It is a long-lived, some specimens in Africa and India and grown over 200 years old. While the tamarind thrives best in deep soil, it flourishes even in rocky terrain with little or no cultural attention. Since it withstands salt-spray, it can be planted fairly close to the seashore regions.

Production in India is mainly concentrated in the drier southern states and the produce is collected by villagers and sold in the open market. Tamarind is not grown on a plantation scale but trees in patches are common in the villages in many states. In some parts of India, it naturally regenerates on wastelands and forest lands. In India it has naturalized in the South India with the forest types of moist deciduous, dry deciduous and tropical dry evergreen forest. Recently, it has been introduced along with other species in enrichment planting in the dry deciduous and thorn forests of the Southern States. It is a useful species for plantation programmes under social forestry.

1.2. CLIMATE AND EDAPHIC FACTORS

Tamarind is adapted to a wide range of ecological conditions by reflecting its wide geographical distribution in the sub and semi-arid tropics.

1.2.1. Climate

(a) Water

Tamarind grows well with an evenly distributed mean annual rainfall of 500 - 1500 mm and with the minimum rainfall requirement for the tree to survive is 250 mm. The maximum annual rainfall which tamarind can tolerate is up to 4000 mm, provided that the soil is well-drained. Tamarind grows under these conditions in the wet tropics but does not flower. In fact, dry weather is important for flower initiation and if heavy rains occur during flowering tamarind does not bear fruit. Dry weather is very important for fruit development and regardless of total annual rainfall; tamarind produces more fruit when subjected to a fairly long annual dry period. Tamarind can be grown in very dry areas with supplementary irrigation and can withstand up to six months without rainfall; this is due to its deep and extensive root system. In the intermediate zone above 300 mean sea level, a few trees can be seen, but their bearing ability is comparatively less than those grown in the dry areas. Tamarind is usually evergreen but may shed its leaves in very dry conditions during the hot/dry season.

(b) Temperature

Essentially a tree of tropical climates, tamarind thrives under a maximum annual temperature ranging from 33-37°C to a minimum of 9.5-20°C. The trees are very sensitive to fire and frost, which requires protection during earlier stage of growth (seedling and sapling condition). Older trees are more resistant to extreme temperature than young trees and can withstand temperatures as high as 47°C and as low as -3°C without serious injury.

(c) Light

Tamarind is a light demanding tree and grows very slowly. It is often isolated from other vegetation and the ground is usually bare around the tree due in part to the dense shading by the canopy.

(d) Wind

The tamarind tree is very resistant to strong winds and can tolerate violent typhoons and cyclones. Often known as the hurricane-resistant tree, it has strong and pliant branches with deep and extensive root system, which anchors it to the ground. Another explanation for the isolated growing habit of the tree, which is tolerance to harsh environments.

(e) Altitude

Tamarind has been reported to grow from altitudes of 0-2000 mean sea level. The lack of tamarind growing at higher altitudes is associated more with the decrease in temperature than the altitude itself.

1.2.2. Soil

The tamarind tree can grow in a wide range of soils to have no specific soil requirement, but no cultivation can flourish in poor soils and on rocky terrain. In India, it tolerates sodic and saline soils where it grows in ravines and on degraded land. Older plants are more resistant to salinity than seedlings and have been found growing close to the sea. However, tamarind thrives best in loamy, deep, well drained alluvial soil, which favours the development of a long tap root (Gupta and Mohan, 1990). Silt-clay and clay soils were not good for plantations in Mali, but grow well in red sandy loam soils at AP, India. The tree does not tolerate under water-logging area. The optimum pH for tamarind is 5.5-6.8, which is slightly acidic, though it grows well in alkaline soils (Fig. 1.2).

Fig. 1.2. Soil Collection in Tamarind Plantation at Dharmapuri District of Tamil Nadu

1.3. BOTANICAL DESCRIPTION

Tamarind is a long-lived, large, evergreen or semi-evergreen tree with 20-30 m tall, 1.5-2 m thick trunk and 8 m in circumference. The trunk forks at about 1 m above ground and is often multi-stemmed with branches widely spreading, drooping at the ends and often crooked but forming a spreading, rounded crown.

1.3.1. Leaves

Leaves are alternate and even pinnate, in length of 7-12 cm, shortly petiolated (up to 1.5 cm long) and petiole glabrous or puberulent as is the leaf rachis. Laminae are glabrous or puberulent, glaucous underneath and darker green above. Venation is reticulate and the midrib of each leaflet is conspicuous above and below.

Leaflets are 10-18 pairs/leaf, each narrowly oblong, rounded at the apex and slightly notched and asymmetric with a tuft of yellow hairs; at the base obliquely obtuse or subtruncate. At the leaf base is a pulvinus and two small stipules 0.5-

1.0 cm long which are caducous early on with stipules are falcate, acuminate and pubescent. A permanent scar is seen after leaf fall (Fig. 1.3).

Fig. 1.3. Tamarind Tree - Leaves

1.3.2. Flowers

Flowers are borne in lax racemes, which are few to several flowered (up to 18), borne at the ends of branches and shorter than the leaves with the lateral flowers are drooping. Flowers are irregular 1.5 cm long and 2-2.5 cm in diameter each with a pedicel about 6 mm long, nodose and jointed at the apex. Bracts are ovate-oblong, and early caducous, each bract almost as long as the flower bud. There are 2 bracteoles, boat shaped, 8 mm long and reddish. The calyx is 10 mm long with a narrow tube (turbinate) and 4 sepals, unequal, ovate, imbricate, membranous and coloured cream, pale yellow or pink. Corolla of 5 petals, the 2 anterior reduced to bristles hidden at the base of the staminal tube. The 3 upper ones are a little longer than the sepals, 1 posterior and 2 lateral, these 3 obovate to oblong, imbricate, coloured pale yellow, cream, pink or white, streaked with red (Fig. 1.4).

Flowers are bisexual. The colour of the flowers is the same on each tree; they are not mixed. Among 4 stamens, 3 are fertile and 1 is sterile. Filaments of fertile stamens are connate and alternate with 6 brittle-like staminodes. Stamens are united below into a sheath open on the upper side and inserted on the anterior part of the mouth of the calyx tube. Anthers are transverse, reddish brown and dehisce longitudinally. The ovary is superior with few too many (up to 18) ovules. The ovary is borne on a sheath adnate to the posterior part of the calyx tube. It is stipitate, curving upwards and is green with a long hooked style with a terminal subcapitate stigma.

Fig. 1.4. Tamarind Tree - Flowers

1.3.3. Bark

The bark is brownish-grey, rough and scaly. Young twigs are slender and puberulent. A dark red gum exudes from trunk and branches, when they are damaged (Fig. 1.5).

Fig. 1.5. Tamarind Tree - Bark

1.3.4. Fruits

The fruits are 5-10 cm long x 2 cm broad, oblong, curved or straight, with rounded ends, compressed and indehiscent although brittle. The pod has an outer epicarp which is light grey or brown and scaly. Within is the firm but soft pulp which is thick and blackish brown. The pulp is traversed by formed seed cavities, which contain the seeds. Each pod contains 1-12 seeds which are flattened, glossy, orbicular to rhomboid, each 3-10 x 1.3 cm and the centre of each flat side of the seed marked with a large central depression (Fig. 1.6).

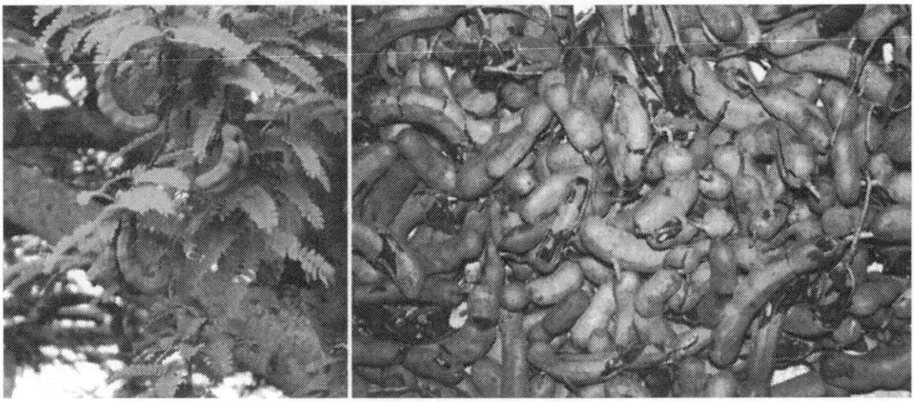

Fig. 1.6. Tamarind Fruits

1.3.5. Seeds

Seeds are hard, red to purple brown, non arillate and exalbuminous. Seed chambers are lined with a parchment like membrane. Cotyledons are thick (Fig. 1.7).

Fig. 1.7. Tamarind Seeds

1.3.6. Root

Tamarind produces a deep tap root with an extensive lateral root system, but the tap root may be stunted in badly drained or compacted soils. The tap root is flexuous and lateral roots are produced from the main root at different levels.

1.4. REPRODUCTIVE BIOLOGY

Natural or unselected tamarind is a long-lived tree and once it starts bearing, its fruit yielding capacity increases with age. The productive life of a tree can last for 50-70 years after which it declines. The normal lifespan of the tree is approximately 150 years. It is important to understand the variation in time of the juvenile phase with genotypic and phenotypic variations in flowering, fruit setting and productivity so that domestication can be enhanced.

1.4.1. Flowering and fruiting phenology

In tamarind, terminal vegetative shoots which bear flowers only in the following flowering season are produced annually. Two types of terminal shoots have been observed, short ones with an erect habit and long ones with a drooping habit. This has been identified as a useful character to evaluate genotypes, since terminal shoot length and foliage production are found to be highly correlated. In an inflorescence the flowers only open on alternative days. Flowering and fruiting times vary from country to country and region to region depending on the latitude and altitude. It is reported that production of flowers varies considerably between selections in India. Those with longer vegetative terminal shoots produce more flowers.

Flowers emerge on new shoots produced in the spring or summer in most countries. However, some trees flower as late as July or August in the rainy season. This common phenomenon is a problem in orchard management, particularly orchards of sweet tamarind. In India early, mid and late flowering types of tamarind have been identified and those with delayed flowering habit are reported to be high yielders.

The period from flowering to pod ripening is 8-10 months. Ripe fruits, however, may remain on the tree until the next flowering period. In most of the tamarind producing countries the fruits are harvested from February to March/April, but sometimes the harvesting period may extend to June. In India, fruits are harvested from April to May although in Kerala and other parts of South India fruit collection may be over by the end of February.

1.4.2. Pollen morphology and pollination vectors

Pollen grains are dimorphic, radially symmetrical, tricolporate, oblatespheroidal in shape and sticky. The pollen dimorphism in tamarind with two distinct sizes of pollen grains: 40-42 μm and 22-25 μm. This size variation was attributed to resource limitation towards the latter part of the flowering season. Smaller pollen grain counts increased during the later part of the flowering. Pollen sterility is very low in tamarind (less than 2%). Under ambient conditions (37-40°C) pollen viability was observed to be 88 per cent for three days, while pollen stored at 40°C remained 97 per cent viable up to 100 days.

1.4.3. Fruit set, breeding and mating system

In spite of profuse flowering, fruit set in tamarind is very low under open pollination. In Tamarind about 10-15 per cent of flowers developed as fruits. This may be due to the short-lived nature of the flowers (about 48 hours) and also due to pollinator limitation, thus many flowers appear not to be pollinated during their short stigmatic receptive period. All unpollinated flowers drop within two days.

In contrast, the controlled cross pollination fruit set was more than 75 per cent, whereas controlled self-pollination resulted in 2-6 per cent fruit set. The controlled cross pollination results in higher fruit set and retention in tamarind than open or self pollinated. Fruit set was only 36 per cent with open pollination whereas it increased to 56 per cent with cross-pollination.

Tamarind is a predominately outcrossing species with some level of selfing. This may be the reason for bearing of fruits even in isolated trees.

1.4.4. Fruit development

Fruit development in tamarind has three distinct stages: growth, maturation and ripening. The indehiscent pods ripen about 8-10 months after flowering and may remain on the tree until the next flowering period. The fruits begin to dehydrate 203 days after fruit set and continue until 245 days when full ripeness occurs and up to half the fruit's original water content is lost.

The fully ripened fruits can remain on the tree for six months. Fruits are harvested at two stages depending upon the market. Green pods are harvested for flavouring and ripe pods for processing.

The fruits of sweet types are also harvested at two stages, half ripe and fully ripe. At the half ripe stage the pulp is yellowish green and has the consistency of an apple, particularly in the case of sweet forms. At the fully ripe stage the pulp shrinks, due to loss of moisture, and changes to reddish brown and becomes sticky.

1.5. PROPAGATION TECHNIQUES

The methods for propagating the Tamarind trees are

1. Seed propagation
2. Vegetative propagation

3. Micro propagation

The methods reviewed here indicate that propagation will not be a constraint to the production of quality planting material for those interested in establishing tamarind orchards.

1.5.1. Seed propagation

Propagation by seed is the least expensive means of propagation and can be used to produce a large number of new plants from a minimum stock material. In addition, seed propagation is necessary as rootstock, which is used for grafting the selected cultivars. The disadvantage of seed propagation is that seeds fail to produce true to type and also result in taking a longer time yield of fruit than those produced through vegetative propagation.

(a) Seed Collection

Seed collection for propagation involves selection of vigorous, disease-free mother trees of superior phenotypes from forests, pastures or farms. Seed viability and storage life are greatest if seeds are collected when mature. Fruits are collected only when they are fresh and ripe as indicated by a buff, brittle epicarp. Pods are shaken from the tree or if reachable fruit pedicels are clipped. The pods should be dried in the sun and the seeds removed from the pulp by hand-kneading, or by soaking in water for several hours. Washed seeds are then dried in the shade and stored in well-ventilated gunny bags in cool place. Total number of seeds on an average of 1800 to 2600 per kg.

(b) Seed Storage

The seed storage behaviour of tamarind is orthodox. Fresh seeds retain viability for atleast six months when kept dry at ambient temperature. Seeds store well in dry sand, remaining viable for at least two years if kept in closed vessels for protection from insect pests. Reducing the temperature is also helpful in maintaining viability of dry seeds. The seeds could be stored for several years in air-tight packs at 10°C and 7-15 per cent moisture content.

(c) Seed treatments

Germination of tamarind seed can be accelerated by overcoming the hard seededness through acid or mechanical scarification and soaking in hot water. Scarification with 97 per cent concentrated sulphuric acid for at least 45 minutes at 25°C has proved to be highly effective with tamarind seed. The seed size and duration of acid scarification are positively related, where bigger sized seeds require 20 minutes of acid scarification while only 15 and 10 minutes respectively was needed for medium and small sized ones (Parameswari *et al.*, 2001).

The hard seed coat of tamarind can also be softened by hot water treatment to overcome seed coat dormancy and stimulate germination. The germination response of tamarind seed to hot water treatment increases with increasing temperature (up to 100ºC) and soaking period (up to 60 min). Mechanical scarification has also been effective in breaking the seed coat dormancy. Scratching the region near the hilum on coarse sand paper or nicking it with a knife increased the rate of germination.

(d) Seed Sowing

Under natural conditions the pods fall from the trees and the seeds are released by decomposition of the pods and these seeds germinate if conditions are favourable. Germinated seedlings often survive under the parent trees if protected from grazing animals. Under artificial condition, seed bed should be prepared for germinating the seed. A normal nursery potting mixture containing three parts of top soil, one part of sand and one part of compost can be successfully used for germinating tamarind seed. In nursery beds, the recommended spacing is 20-25 cm both ways. The best germination results if the seeds are covered by 1.5 cm of loose sandy loam or a mixture of loam and sand. Adequate moisture is needed to ensure germination and seedling establishment is as uniform as possible.

(e) Germination

Germination of tamarind seed is epigeal. The radicle swells and emerges from one end of the seed and descends rapidly. The hypocotyl elongates arches slightly and raises the cotyledons and shoot apex above the ground towards the light. On average, tamarind seeds begin to germinate about 13 days after sowing, but take a month to complete. The germination pattern is similar to that of other leguminous species in which the hard seed coat interferes with water and oxygen uptake, thereby limiting germination. The germination capacity of fresh or well stored tamarind seed is reported to vary from 65 per cent - 75 per cent (Fig. 1.8).

Fig. 1.8. Tamarind Seedlings

Seedlings grow rapidly in the early stages and produce a long tap root which may reach 30 cm or more within two months of germination. Seedlings should attain a height of at least 80 cm before being transplanted to the field at the beginning of the rainy season.

1.5.2. Vegetative Propagation

(a) Cuttings

Tamarind coppices well. These can be removed and used for small scale propagation. However, the easiest and the cheapest method of vegetatively propagating tamarind are by stem cuttings. Cuttings of shoots bearing new flushes of fully turgid leaves are collected in the morning. They are immediately dipped for ten seconds in 1000 ppm of indole butyric acid (IBA), and in 50 per cent isopropyl alcohol, before being planted in polypropylene tubes containing vermiculite/perlite (1:1) and placed in a mist propagator with 70-80 per cent humidity (Fig. 1.9).

Fig. 1.9. Vegetative propagation through cuttings (taken from coppices)

The use of the growth regulator namely indole butyric acid increased the rooting of cuttings to over 94 per cent when compared to the control, which recorded only 25 per cent. Furthermore, the time taken to initiate rooting was only 10-15 days with indole butyric acid, while the control took 40-50 days. Terminal cuttings have an advantage over mid stem cuttings because there is only one cut end thus reducing the possibility of infection by disease causing organisms during or after the rooting phase.

Soft or semi-soft stem cuttings (15-20 cm long), taken from 1-2 year old branches can also be rooted. In this method, the cuttings are wrapped in moist cloth after removal from the trees to prevent moisture loss. They are then dipped in a rooting hormone, preferably IBA at 1000 ppm and placed in a sand bed in a mist chamber. Bud initiation and root formation occurs after 20 days and new leaves are formed after 45 days. Hard wood cuttings will not root and hence their use should not be attempted (Reddy *et al.,* 2000).

(b) Budding and Grafting

Vegetative propagation by shield and patch budding, cleft grafting, whip grafting and approach grafting are reliable methods, although they are expensive and time consuming. Successful budding requires that the scion material has fully-formed, mature, dormant buds and that the rootstock be in a condition of active growth such that the 'bark is slipping'. In tropical areas, the bark slips anytime the plant is in active growth, which is practically year-round. For patch budding, seedling rootstocks should be grown in raised beds and budded when they are 6-9 months old. This is a suitable method for large-scale multiplication of Tamarind. The success percent for modified ring budding and patch budding of 9-month-old seedling rootstocks was 94 per cent and 96 percent success respectively (Pathak *et al*., 1992). Shield budding, patch budding and cleft grafting have been used for producing plants for large plantations.

Approach grafting is successful for difficult-to-graft species, even when others techniques fail, because both the scion (upper portion of the grafted pair) as well as the seedling root stock remain attached to their own root systems during the period of graft union formation. Approach grafting has been used in tamarind with up to 95 per cent success. The veneer grafting method possesses promise for mass scale commercial propagation. For conducting this grafting operation, rootstocks about six months old and of uniform size should be selected. The scions should be defoliated before grafting. Veneer grafts are made eight cm high on the root stock and immediately after inserting the scion the root stock is removed above the graft union.

For softwood grafting, the rootstock seedlings should be defoliated and their tops cut off at 15 cm high immediately before grafting. A vertical downward cut is made in the centre of the cut stem to about four cm depth and the scion sticks are cut into wedge shapes, inserted into the stock and wrapped using two cm wide 200 gauge polyethylene ribbon. Soft wood grafting has been shown to be the best grafting method in terms of successful unions and survival rates.

Naturally defoliated, vigorous and terminal scions are disinfected with fungicide and used the same day they are excised. Splice side grafting resulted in 79 per cent success after 20 days. In the sub-humid parts of India, experiments revealed that the time of grafting significantly influenced per cent sprouting, days to sprout, per cent graft success, and linear and radial growth. Softwood grafting in April result in highest graft success (83%), maximum linear growth (56 cm and maximum saleable plants (74%) in the following March. Under the conditions pertaining in Karnataka, India, the time required for grafts to take was longer during the winter months (October to December) than during other months (Bharad *et al.*, 1999). As the temperature increases, the time taken for graft union and sprouting decreases, with the least number of days (14) being observed in March. The highest success percentage (100%) in cured scions was recorded during March, closely followed in February (94%) by Reddy *et al.* (2000).

Fig. 1.10. Vegetative propagation through approach grafting

The correct spacing of mother plants is important to produce shoots for commercial production of grafted plants. The age of the rootstocks for soft wood grafting is also important for success. Scions pre-conditioned for 30 days prior to grafting on to 6 and 9 month-old rootstocks growing in the green house resulted in 69-72 per cent success at 60 days. Grafting success has been attributed to the fact that rootstocks of this age contain a higher proportion of reducing sugars to total sugars than at other ages. Provenance may also influence traits such as hardiness, growth rate and adaptability to climatic conditions, and this should be kept in mind when purchasing or collecting scion materials from different locations. Root stock selection for vegetative propagation of tamarind is important as it controls the vigour and the equilibrium between yield and quality. Dwarfing rootstocks are considered best because they result in short trees, which are easy to manipulate for management practices, have increased fruit production, and require less land and labour inputs. A dwarf rootstock has been identified based on morphological characteristics, such as internodes length and leaf area. These characters are correlated to the number of stomata on the leaves (Fig. 1.10).

Attempts to reduce canopy size by intergeneric grafting tamarind scions on to root stocks of other leguminous species have not been successful. The applications of growth inhibiting chemicals have no dwarfing effect.

(c) Layering

Layering is a simple and easy method of rooting a branch of a shrub or tree while it is still attached to the parent plant. It is a widely practiced propagation method for tamarind. In ground layering, a flexible branch is bent down and pegged to the ground, and the point of contact covered with soil. A small cut is made in the lower side of the stem where it touches the ground to impede sap circulation and encourage rooting. After 3-6 weeks roots form at the point of contact with the soil and a new plant can be obtained by severing the branch above the place of rooting.

For air layering or marcotting, a branch is selected and a 2.5 cm length of the bark is removed. This area is covered with a moist soil mixture or a moist porous material, such as coir fibre dust, and kept in position by wrapping with polyethylene film. It is kept moist for 2-3 months. When large quantities of roots are observed through the polyethylene film, the branch with its roots is severed from the parent tree and potted up for later planting.

Girdling and etiolation increases the percentage of total sugars, reducing and non-reducing sugars, lowers the starch content and raises the percentages of soluble and total nitrogen. This result in a lower C:N ratio, a higher level of the rooting cofactor and lower levels of indigenous inhibitors compared with nongirdled etiolated shoots. The application of indole butyric acid (IBA) or IBA + NAA (naphthalene acetic acid) increases the number and the length of air-layered roots, and reduces the time for rooting from about 10-6 weeks.

1.5.3. Micro-propagation

Tamarind may be propagated by tissue culture techniques; the report on in vitro morphogenesis in this species is limited, due to its recalcitrant and callogenic nature. To overcome these limitations, attempts have been made to induce meristematic activity in seedling explants. Seedlings are germinated in media with or without thidiazuron (TDZ). Presence of this growth regulator restricts the differentiation of the apical meristem to form shoots.

It triggered proliferation of the meristematic tissue at the cotyledonary node and a large number of meristematic buds appear in a radial pattern around the node. The meristematic activity extended to the junction of the epicotyl and hypocotyl, giving rise to buds in the form of protuberances from all sides of the junction.

1.6. MAJOR TAMARIND VARIETIES

- PKM 1
- Urigam
- Hasanur
- Tumkur
- Prathisthan
- DTS 1
- Yogeshwari
- Red Tamarind
- Sweet Tamarind
- Harur

1.7. PRODUCTION OF TAMARIND IN DIFFERENT STATES OF INDIA

1.7.1. State-wise Area and Production of Tamarind in India

Tamarind production in different states of India was assessed and presented in Table 1.1. The initial survey was obtained from 7 states and 1 Union Territory namely Andhra Pradesh, Karnataka, Kerala, Maharashtra, Puducherry, Tamil Nadu, Chhattisgarh and Tripura. Comparing the area wise production, Tamil Nadu contributed the maximum area with the 21,11,000 ha. followed by Karnataka (16,80,000 ha) and minimum cultivation area of 60.00 ha. was observed in Puducherry. Whereas in fruit production, the highest production was recorded in Karnataka (87.00 MT) followed by Tamil Nadu (67.52 MT) and minimum was recorded in Puducherry with the production of 0.35 MT.

Table 1.1. Area and Production of Tamarind in different states

Sl. No	States	Area in ('000 Ha)	Production in ('000 MT)
1	Andhra Pradesh	5.53	20.00
2	Karnataka	16.80	87.00
3	Kerala	14.00	20.00
4	Maharashtra	5.70	11.40
5	Puducherry	0.06	0.35
6	Tamil Nadu	21.11	67.52
7	Chattisgarh	9.25	1.20
8	Tripura	0.087	0.263

(Horticulture Statistics, 2014)

1.7.2. District-wise Area and Production of Tamarind in Tamil Nadu

The district wise details regarding the area and production of the Tamarind fruit was tabulated (Table 1.2).

Table 1.2. District Wise Area and Production of Tamarind in Tamil Nadu

District	Area (In Hect.)	Production (Tonnes)
Ariyalur	250	653
Coimbatore	855	5194
Cuddalore	218	1052
Dharmapuri	867	2760
Dindigul	4467	13464
Erode	403	2374
Kancheepuram	52	209
Kanyakumari	1610	3131
Karur	111	302
Krishnagiri	1114	3874
Madurai	1410	4255
Nagapattinam	288	655
Namakkal	305	1165
Perambalur	384	1043
Pudukottai	324	803
Ramanathapuram	273	904
Salem	282	1248
Sivagangai	423	1507
Thanjavur	192	446
The Nilgiris	0	0
Theni	2838	8318
Thiruppur	180	720
Tiruchirapalli	1381	3178
Thirunelveli	841	1933
Thiruvallur	48	182
Thiruvannamalai	62	344

District	Area (In Hect.)	Production (Tonnes)
Thiruvarur	136	321
Thoothukudi	427	997
Vellore	737	4300
Villupuram	181	814
Virudhunagar	455	1431
Total	21114	67524

(Horticulture Statistics, 2014)

Among the 32 districts surveyed, the maximum tamarind area cultivated is Dindigul (4467 Ha.) followed by Theni (2838 Ha.), Kanyakumari (1610 Ha.), Madurai (1410 Ha.), Tiruchirapalli (1318 Ha.), Krishnagiri (1114 Ha.), Dharmapuri (867 Ha.), Coimbatore (1318 Ha.) and the minimum tamarind area is 0 Ha. (The Nilgiris).

Whereas in the production was expressed in Tonnes. The maximum tamarind fruit production was observed in Dindigul (13464 Tonnes) followed by Theni (8318 Tonnes), Coimbatore (5194 Tonnes), Vellore (4300 Tonnes), Madurai (4255 Tonnes), Tiruchirapalli (3178 Tonnes) and the minimum production in The Nilgiris with the value of Nil (0 Tonnes).

REFERENCES

Bharad, S.G., Gholap, S.V., Dod, V.N. and Bople, S.R. (1999) Seasonal variation in success of softwood grafting of Tamarind (*Tamarindus indica* L.). *Journal of Applied Horticulture,* **1(1)**: 57-58

El-Siddig, Gunasena, Prasad, Pushpakumara, Ramana and Vijayanand, Williams (2006) Tamarind (*Tamarindus indica*). Southampton Centre for Underutilised Crops, Southampton, UK

Gupta, G.N. and Mohan, S. (1990) Response of several tree species to management on degraded soils of semi arid region. *Indian Forester,* **116(8)**: 622-630

Handbook on Horticulture Statistics (2014) Department of Agriculture and Cooperation, Ministry of Agriculture, Government of India, New Delhi

Parameswari, K., Srimathi P. Malarkodi K. (2001) Influence of seed size and duration of acid scarification on seed germination of Tamarind (*Tamarindus indica* Linn.). *Madras Agricultural Journal,* **88(1/3)**: 56-60.

Pathak, R.K., Ojha C.M. and Dwivedi, R. (1992) Adopt patch budding for quicker multiplication in Tamarind. *Indian Horticulture,* **36(2)**: 17.

Reddy, B.S., Patil, D.R., Patil, H.B. and Thamiah N. (2000) Effect of time on success of detached grafting in tamarind. *Advances in Agricultural Research in India,* **10**: 141-143.

■■■

Chapter **2**

CULTIVATION AND PRODUCTION

Dr. C.N. Hari Prasath*, Dr. P. Pretheep Kumar &
Dr. I. Jaisankar*****

**Senior Research Fellow, Department of Silviculture,
Forest College and Research Institute, Mettupalayam
**Assistant Professor (Agrl. Entomology), Department of Silviculture,
Forest College and research institute, Mettupalayam
***Senior Scientist, Department of Horticulture and Forestry,
Central Island Agriculture Research Institute, Andaman & Nicobar Islands*

2.1. FIELD ESTABLISHMENT

Tamarind is grown within various cropping systems. The trees is grown as an orchard crop in a pure stand, as an agroforestry species in mixed cropping systems including home gardens or hedgerow tree.

2.1.1. Direct Seeding

Tamarind is established in the field by directly sowing seed. Seeds may be sown directly to establish plantations, hedgerows, or home gardens. Once the site is properly prepared, seeds should be planted as soon after collection as possible to minimise loss of viability. However, seed which has been stored under optimal conditions may be used. The size of the planting hole is of little importance, but firm seed-to-soil contact and coverage is required for adequate germination. Planting holes should have been previously prepared and filled with well-decomposed manure or compost. The seed should be planted no deeper than 1.5 cm. For planting on the square, on the triangle or in lines, a few seeds are dibbled in 5 cm apart at each planting station. Planting stations should be approximately 4-5 m apart. This method is practised in the Indian states of Tamil Nadu and Andhra Pradesh. After germination, when the seedlings reach a height of 10-20 cm, they should be thinned to one at each planting point and the best seedlings being selected.

2.1.2. Transplanting

Transplanting nursery grown plants into the field have advantages over direct seeding as the plants are already established. However, most newly planted trees are subject to 'transplant shock', which may result in increased vulnerability to drought, insects, diseases and other problems. To a greater or lesser degree, transplant shock lasts until the natural balance between the root system and the top or crown of the transplanted tree is restored. The chance of survival can be significantly improved by using practices which favour the development of the root system. Regular care during the first 2-3 years following transplanting is thus very important. In order to provide space and an ideal medium for the development of a vigorous and deep tap root system planting should be done in 1 x 1 x 1m pits filled with well rotted organic manure at the time of planting.

Nursery grown seedlings or vegetatively propagated plants can be raised in nursery beds and delivered to the field as bare rooted plants. Alternatively they can be grown in, or transplanted into polyethylene bags and taken to the field as container grown plants. Nursery-grown trees are usually transplanted during the early rainy season.

Nursery produced plants can be transplanted to the field at about 12-14 months, by which time they are about 80 cm tall. Under optimum growing conditions they may achieve this height in 4-6 months. In India, seedlings ranging from 0.4-2 m in height are commonly planted. If growth is poor, the seedlings should be retained in the nursery for another year. However, it is difficult to handle these older seedlings as their tap roots may have penetrated deep into the soil, in which they will require root pruning and care should be exercised in transporting them. Overgrown seedlings can be more effectively transplanted as stumped seedlings, with stem and tap root pruned to lengths of about 5 cm and 20-25 cm, respectively.

2.1.3. Field spacing

The spacing between plants may range from 5-13 m apart. Spacing may be even wider on fertile soils where the trees grow larger. When establishing a pure stand plantation, the final spacing should be at least 13 x 13 m square or on the triangle. Inter-cropping of young trees with a range of annual crops is possible at this spacing There can be an advantage in planting at higher density (4 x 4 m or 5 x 5 m), followed by thinning two or three times to the appropriate final spacing of 8-15 x 8-15 m. This allows some further selection in the field and reduces the weeding costs in the early years. In parts of India, tamarind is established at 8 x 8 m, 8 x 12 m or 12 x 12 m.

The final spacing can be closer when vegetatively propagated plants are used, as they do not attain the height or spread like seed propagated trees, and their management is easier. A few promising early-bearing varieties with high yield potential and attractive fruit qualities are now available in India. Some of the most popular cultivars in the south are PKM - 1, Pradishtan and Urigam. Grafted plants of these cultivars are being grown in Tamil Nadu either as isolated trees or in commercially viable groves in dry tracts.

A spacing of 10 x 5 m has been recommended for commercial planting, and about 250 trees will be enough to cover a hectare. In India, tamarind is used to make firebreaks, as the soil beneath is almost bare. When creating firebreaks, the spacing adopted is 2.5 x 2.5 m or 3 x 3 m. If the spacing appears to be too close some plants can be removed later. Aftercare operation is minimal for these plantings, the only requirement being to hoe the soil around the plants once a year to remove any weed growth.

2.1.4. Time of Planting

In seasonally dry regions, the best time for field planting is at the beginning of the rainy season, as soon as there is sufficient moisture in the soil. This will reduce the need for frequent watering until the plants are firmly established in the soil. In the initial stages of growth, the plants may need some watering, especially during the hot summer months. If irrigation is available, field establishment may be undertaken at any time of the year, even in the dry season. However, it is advisable to provide partial shade to the newly established plants if planting is carried out during the dry periods (Fig. 2.1).

Seedling growth in the field is initially fast (about 1.2 m in the first two years) but slows later. Mean plant height increases by about 60 cm annually. Average annual growth has been reported as 50-80 cm. However, when bare rooted seedlings are transplanted initial growth may be severely checked. This can be overcome by removing some of the leaves and irrigating frequently. Transplanted trees should be protected from cattle and other animals. Regular weeding around the plants is also essential for early establishment and good growth.

Fig. 2.1. Tamarind seedling planted in research field

2.2. SILVICULTURAL MANAGEMENT

In its wild state, tamarinds establish through natural regeneration of seeds, grow slowly without any addition of fertilizers and starts bearing fruits after a long period and produce fruits usually of inferior quality. Major constraints to the development and promotion of tamarind include low yielding and poor quality planting stock, lack of production technologies and lack of knowledge on nutrition management of the trees. Nevertheless, in recent years tamarind has attracted considerable attention from research and development agencies in many countries.

Young plants are very frost tender and need protection for several years. Individual plants can be protected by covering with a cape made of grass. Though the seedlings stand drought better than frost it is advisable to water the plants for first 2-3 years. They are very responsive to weeding and soil-working. The young seedlings may be suppressed back and killed by the weeds. Seedlings also need protection from grazing animals. The directly sown crops need thinning out in the subsequent years.

It is a slow growing species, but it is grown for valuable fruits. The tree has high longevity upto 200 years or more. The trees are generally retained in avenues and forests worked on physical rotation.

2.2.1. Intercropping

Tamarind allows intercropping with a variety of annual crops. Vegetables and legumes can be grown during the rainy season in the interspaces in the first three to six years to augment farm income and improve soil fertility. Once fully established, the trees will provide a regular income to the grower even when other annual crops fail at times of protracted drought.

The intercropping period is usually limited to four years for vegetatively propagated trees. If intercrops are not used, cover crops are useful to control competition from weeds and to conserve moisture. Intercrops should be fertilized in their own right to reduce competition for nutrients and to maximize their yield. There is a wide choice of intercrops which can be grown depending on the soil, climate and local market demand.

2.2.2. Pruning and training

Initial training and pruning of young plants during the first years is essential for the development of well-formed trees. Tamarind is a compact tree and produces symmetrical branches. Young trees should be pruned to allow 3-5 well-spaced branches to develop into the main scaffold structure of the tree (Fig. 2.2).

Bearing trees require very little pruning other than maintenance pruning to remove dead, weak and diseased branches and water sprouts. In closely planted orchards regular pruning is needed to rejuvenate fruiting wood and control the size of the trees.

Fig. 2.2. Canopy management in the Tamarind plantation for inducing the flowering

Fig. 2.3. Irrigation on Tamarind plantation

2.2.3. Irrigation

Irrigation is not normally practised in tamarind cultivation, but promotes better growth during establishment and the early stages of growth, especially during the dry seasons. Where irrigation facilities are available watering should be done and repeated as the need arises in the early stages of growth. In later years as the deep tap root system develops, the need for watering becomes less. Flowering and fruiting is promoted by irrigation. In dry areas, the use of water harvesting techniques during the rainy season should be considered as it encourages subsequent growth and fruiting (Fig. 2.3).

Mulching during the dry season will also help to reduce water losses from evaporation. Mulches around the trees also help in weed control and water conservation.

2.3. SOIL AND NUTRIENT MANAGEMENT

2.3.1. Soil

The tamarind tree can grow in a wide range of soils and has been suggested by to have no specific soil requirement. With little or no cultivation it can flourish in poor soils and on rocky terrain. In India, it tolerates sodic and saline soils where it grows in ravines and on degraded land. Older plants are more resistant to salinity than seedlings and have been found growing close to the sea. In fact, in northeast Thailand, tamarind has been reported to establish naturally in areas with recently salinised soils. Tamarind could grow in soil containing up to 45 per cent exchangeable sodium and seedling establish in saline media. However, tamarind thrives best in loamy, deep, well drained alluvial soil, which favours the development of a long tap root. The tree does not tolerate water-logging it prefers for well aerated soils. The optimum pH for tamarind is 5.5-6.8, which is slightly acidic (FAO, 1988), though it also grows well in alkaline soils.

2.3.2. Nutrition

2.3.2.1. *Fertilizers and manures*

Trees are known to fruit well even without fertilizer application, due to their deep and extensive root system. In experimental plots in India, 25 g urea and 25 g single super phosphate per plant have been used, although response of tamarind to fertiliser application is poor (Gupta and Mohan, 1990). In India, inorganic fertilizers are not normally applied to tamarind trees, but 5 kg of farmyard manure is applied to the planting hole at the time of planting. Every year thereafter 5 kg of farmyard manure and 5 kg of neem cake are applied per tree in the months of March and April.

The amount of fertiliser is gradually increased as the trees grow. When the tree begins to bear fruits, about 500g of a complete fertiliser containing high amounts of nitrogen and potassium is applied per tree twice a year. A full bearing tree should receive 2-3 kg of a complete fertiliser (Nitrogen, Phosphorus and Potassium) each year, the formula depending upon the nutrient content of the soil.

The foliar spray of Cycocel (1500 ppm), Ethrel (500 ppm), Triacontanol (20 ml $tree^{-1}$), IBA (150 ppm), Planofix (100 ppm), micronutrient mixture (0.5%), $ZnSO_4$ (0.5%) + boric acid (0.3%) + $FeSO_4$ (0.5%) and Urea (1.5%) significantly improved flowering, pod set and retention, yield and quality attributes of tamarind compared to non-sprayed trees. Treatment with foliar feeding urea result in maximum number of flowering (76%), with Cycocel resulted in maximum pod set (32%) and retention (55%) per unit area and with Triacontanol (20 ml $tree^{-1}$) resulted in maximum pod yield (8.3 kg $tree^{-1}$) by Ilango and Vijayalakshmi (2002 a & 2002 b).

Fertilization during the nursery stage is recommended to produce healthy plants for afforestation programmes. Among the treatments, 0.6 g urea, 2 g single superphosphate and 0.6 of muriate of potash with 5 g of compost and 0.5 g of phosphobacteria recorded significant increase in shoot length (48 cm), root length (39 cm) and dry weight (11.2 g).

2.3.2.2. *Nitrogen fixation*

Leguminous species are associated with Rhizobium bacteria in root nodules which fix atmospheric nitrogen. Very little information is available on nitrogen fixing bacteria in tamarind compared to other cultivated legumes. Tamarind was initially considered to be a non-nodulating species. However, some evidence suggests that it does form a symbiotic association with Rhizobium bacteria enabling the tree to fix atmospheric nitrogen under appropriate conditions.

Root nodules collected from tamarind trees grown in plantations on acidic soils as elliptical or circular and light yellow in colour, have been reported to contain high nitrogenase activity. Rhizobial isolates have been described as gram negative, with short and long rods. In the semi-arid region, tamarind has been recognised as a potential nitrogen fixing tree.

The acetylene reduction assay for nitrogen-fixing activity was unable to demonstrate any symbiotic nitrogen fixation in tamarind roots and nitrogen-fixing bacteria were absent in the roots. The most outstanding feature noted in the roots are finger-like projections called 'microvilli,' whose function is unknown, but it has been suggested that they assist in water absorption. Detailed information on the root structure of tamarind is limited.

2.4. PESTS OF TAMARIND

Insect infestation in tamarind is well known for causing direct loss but indirect loss in terms of quality of the produce also impacts its trade and use. Though tamarind is infested by many insect pests, damage by the tamarind beetle assume significance. Tamarind fruits and seeds are highly infested by the Tamarind beetle/Tamarind bruchid during storage which greatly affects the quality of the product. The damage caused by the beetle is particularly significant when the nuts are destined for confectionery purposes.

2.4.1. Tamarind beetle – *Caryedon serratus* – Bruchidae – Coleoptera

Biology

Egg

- Laid singly, glued to the surface of the fruits.
- Fresh eggs are translucent, orange cream in colour, changing to grayish white with age and approximately 1 mm long and 0.5 mm wide (Fig. 2.4).

Fig. 2.4. Tamarind beetle – *Caryedon serratus* egg

Grub

- Fleshy, curved, creamy white in colour with black mouth parts.
- Grub alone cause damage.
- Grub are scarabaeiform and sparsely hairy.
- After hatching, the grub burrows straight through the egg shell and the pod wall, and starts eating the seed. The first sign of attack is the

appearance of 'windows' cut into the pod wall by the grub to allow the adult to leave the pod after emerging from the pupal cocoon.

- Fully grown grub sometimes comes out through the exit holes made by the previous generations. They often live in the storage sacks and pupate in large numbers at the bottom of the pile of sacks (Fig. 2.5).

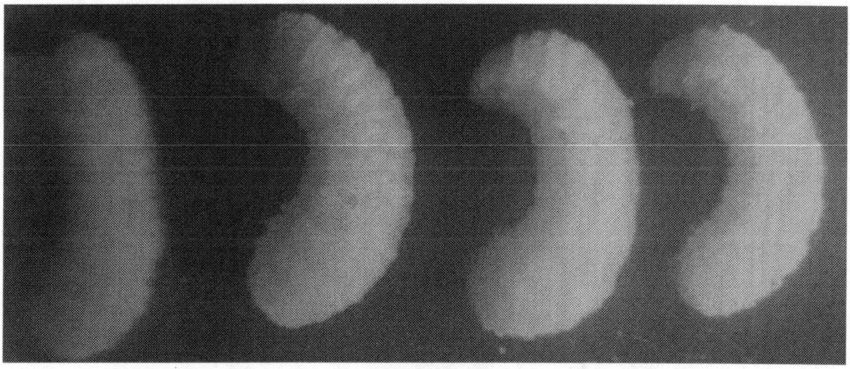

Fig. 2.5. Caryedon serratus grub

Pupa

- Pupation takes place in a pupal cell prepared beneath the seed coat.
- Pupae are creamy white, glabrous, about 5 mm long (Fig. 2.6).

Fig. 2.6. Caryedon serratus pupa

Adult

- Brownish-grey beetle with characteristic elevated ivory like spots near the middle of the dorsal side.
- It is small, short, and active with long conspicuous serrate antenna.

- Head is prominent and pronotum is subconical.
- Serrate antenna.
- Elytra do not cover the abdomen completely, which is called as pygidium.
- Adults are short lived, harmless and do not feed at all (Fig. 2.7).
- Body length is 3.5-7.4 mm.

Fig. 2.7. Caryedon serratus adult

Optimum conditions for development

The optimum conditions for development are 30-33°C and 70-90 per cent RH, under which conditions the development period is 41-42 days.

Symptoms of damage

Circular hole on fruits and seeds of tamarind both in tree and storage (Fig. 2.8).

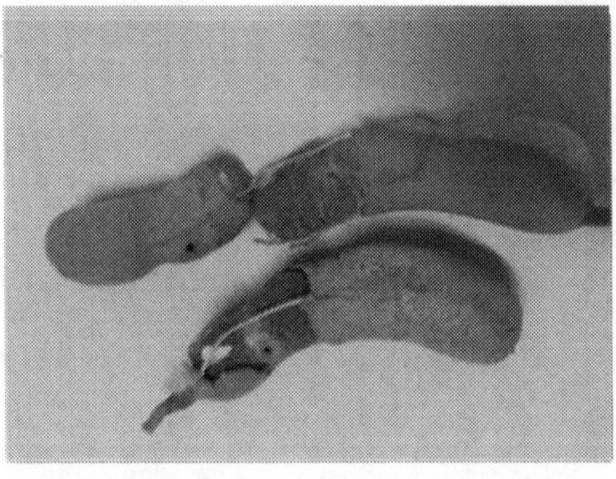

Fig. 2.8. Symptoms on tamarind fruits

2.4.2. Scale – *Aspidiotus tamarindus* – Diaspididae – Hemiptera

Biology

- Yellowish circular scale

Nature of damage

- Nymph and adult cause damage.
- Yellowish encrustation over developing fruit, ill filled and hard fruits.

2.4.3. Soft scale – *Saissetia oleae* – Coccidae – Hemiptera

Biology

- Round greenish brown scale.

Nature of damage

- Nymph and adult cause damage.
- Encrustation over developing fruits.

2.4.4. Blue butterfly – *Virachola isocrates* – Lycaenidae – Lepidoptera

Biology

- Larva is dirty brown in colour.
- Adult is a bluish-brown butterfly with an orange patch on forewing and dark spot on the hindwing.

Nature of damage

- Larva cause damage.
- Bore holes on fruits.

2.4.5. Semilooper – *Achaea janata* – Noctuidae – Lepidoptera

Biology

- Grey coloured larva with lateral red or brown stripes.
- Adult is a pale reddish brown moth with black hindwings having a medial white band and three large white spots on outer margin.

Nature of damage

- Larva cause damage.
- Defoliation.

It has been reported that one of more sprayings with DDT before fruit ripening lowers insect infestation. ***Achaea janata*** can be controlled by spraying Quinalphos 25EC 2ml/lit of Monocrotphos 36 WSC 2ml/lit.

Spraying Quinalphos 25EC 1ml/lit at the time of fruiting season, will protect the tree.

2.4.6. Powdery mildew

Powdery mildew can be controlled by spraying Dinocap 1g /lit.

2.5. TAMARIND RESEARCH IN THE ISLANDS

Being an underutilized fruit crop tree in this Island the research on these species is meager but the perennial leguminous species like tamarind and *Ceiba pentandra* were found generally in each home garden of the Islands. The tamarinds form the third storey (5-10 m) in the system and are closer to the house. Pandey *et al.* (2006) recorded the density of 13 no ha^{-1}, relative frequency of 3.5 per cent and a height of 6 m of tamarind in the home gardens of Andaman and Nicobar Islands.

Albinism

Albinism is governed by single recessive gene and the trait is expressed only when it is in homozygous recessive (aa) condition. Albinism has been reported in many angiosperm and gymnosperm species. To cite a few, the examples are *Pongamia pinnata* (Kumaran *et al.,* 2007), *Simarouba glauca* (Rathakrishnan *et al*., 2003), *Gmelina arborea* (Karoshi *et al*., 2001), *Artocarpus integrifolia* (Rao *et al*., 1999), *Dendrocalamus strictus* (Kumar *et al*., 1993), *Eucalyptus tereticornis* (Venkatesh and Sharma, 1974) and in Slash pine (Squillace and Kraus, 1963). However under natural condition, the frequency of such albino seedlings will vary depending upon the extent of selfing and mating of albino carriers.

In the present study, fifty fruits each of *T. indica* were collected from a single tree per location/place *viz*., Sipighat 1, Sipighat 2, Diglipur and Maya Bunder of Andaman Islands, seeds were extracted from the fruits and sown in the nursery at Sipighat Research Farm, Central Island Agricultural Research Institute, Port Blair (11°60" N, 92°67" E, 22m MSL). The poly bags were filled with Soil: Sand: Farm Yard Manure (2:1:1). The germination of the seed commenced from 8th day after sowing. Out of the four locations, Sipighat 1 showed albinism. Out of 200 seeds sown, 183 germinated, of which nine (4.92%) were albinos

(Fig. 2.8). Among 9 albinos, 5 died 19 days after germination and remaining 4 showed yellowish leaves with malformed growth and survived upto sixth leaflet stage. However, the cotyledons were dark yellow with light green petiole. Since albino seedlings lack chlorophyll the food manufacturing unit, they do not survive for long and die soon after the reserve food material stored in the endosperm cotyledons is exhausted by the seedling (Kumar *et al.*, 1993). The comparison of normal plants and albino with regard to seedling parameters is presented in Table 2.1.

Table 2.1. Comparison of normal and albino seedlings 45 days after sowing

Type of plant	Shoot length (cm)	Root length(cm)	No. of leaflets	Remarks
Green	27.5	25.5	8	Normal growth, green leaves
Albino 1	24.5	8.5	4	Abnormal pale yellowish leaves
Albino 2	20.8	9.4	6	-do-
Albino 3	21.6	8.0	5	-do-
Albino 4	21.3	8.6	4	-do-

In the present case the albinism recorded in the locality Sipighat-1 may be due to one of the species which is carrier for albinism (Aa). Such plants give rise to albino seedlings following selfing or out crossing with albino carriers. The frequency of occurrence of albino seedling will depend on the amount of selfing or out crossing (Fig. 2.9).

Fig. 2.9. Tamarind young seedlings (A – Normal Green, B – Albino)

Obviously, lethal albino homozygote by themselves cannot be of any practical value. However, cloned individuals heterozygous for such mutant traits can serve as useful genetic markers for monitoring the degree of natural crossing and selfing in seed orchards of *Tamarindus indica* as suggested for *Bombax* sp. (Venkatesh and Emmanuel, 1976) and *Gmelina arborea* (Venkatesh *et al.*, 1978).

REFERENCES

FAO (1988) Fruit Bearing Trees. *Technical notes.* FAO-SIDA Forestry Paper 34: 165-167.

Gupta, G.N. and S. Mohan (1990) Response of several tree species to management on degraded soils of semi arid region. *Indian Forester,* **116(8)**: 622-630

Ilango, K. and C. Vijayalakshmi (2002a) Effect of growth regulators and chemicals on pod set and retention in Tamarind (*Tamarindus indica* L.). *My Forest,* **38(2)**: 133-137.

Ilango, K. and C. Vijayalakshmi (2002b) Effect of some growth regulators and chemicals on yield and quality attributes in Tamarind (*Tamarindus indica* L.). *Orissa Journal of Horticulture,* **30(1)**: 35-39.

Karoshi, V.R., G.V. Hedge and S.M. Hiremath (2001). Albino seedlings in *Gmelina arborea* Roxb. *Indian Forester,* **127 (4)**: 477-479.

Kumar, A., V.K. Sharma and B.S. Beniwal (1993). Albino seedlings in *Dendrocalamus strictus Nees. Indian Forester,* **119 (6)**: 507-509.

Kumaran, K., I. Jaisankar, K. Nesamani and M. Paramathma (2007). Albinism in *Pongamia pinnata*(L.) Pierre. *Indian Forester,* **133 (5)**: 709-710.

Pandey, C.B, Kanak Lata, Venkatesh, A. and Medhi, R.P. (2006). Diversity and species structure of home gardens in South Andaman. *Tropical Ecology,* **47(2)**: 251-258.

Rao, P.S., S.S.N. Murti and K. Venkaiah (1999). Albinism in *Artocarpus integrifolia* Linn. F.- A case study. *Indian Forester,* **125(11)**: 1095-1098.

Rathakrishnan,P., P. Rajendran and B.N. Divakara (2003). Albinisn in *Simarouba glauca* Linn. *Indian Forester,* **129 (9)**: 1167-1169.

Squillace, A.E. and J.F. Kraus (1963). The degree of natural selfing in Slash pine as estimated from albino frequencies. *Silvae Genetica,* **12**: 63-50.

Venkatesh, C.S. and C.J.S.K. Emmanuel (1976). Spontaneous chlorophyll mutations in *Bombax* L. *Silvae Genetica,* **25**: 137-139.

Venkatesh, C.S., R.S. Arya and R.C. Thapliyal (1978). An albino-type natural chlorophyll mutant in *Gmelina arborea* Roxb. *Silvae Genetica,* **27:** 40-41.

Venkatesh, C.S. and V.K. Sharma (1974). Some unusual seedlings of Eucalyptus: Their genetic significance and value in breeding. *Silvae Genetica,* **23**: 120- 124.

■■■

Chapter **3**

CHEMISTRY OF TAMARIND

Dr. M. Shakila Banu

Professor and Head, Department of Food Processing and Preservation Technology, Avinashilingam University, Coimbatore

Tamarind is a tropical tree used primarily for its fruits, which are eaten fresh or processed, used as a seasoning. Tamarind fruit pulp has a sweet acidic taste due to a combination of high contents of tartaric acid and reducing sugars. The pulp is used for seasoning, in prepared foods, to flavour confections, curries and sauces, and as a major ingredient in juices and other beverages.

Table 3.1. Nutritive Value per 100 g of edible portion

Constituents	Pulp (ripe)	Leaves (young)	Flowers
Calories	115	-	-
Moisture (g)	28.2-52.0	70.5	80.0
Protein (g)	3.10	5.8	0.45
Fat (g)	0.1	2.1	1.54
Fibre (g)	5.6	1.9	1.5
Carbohydrates (g)	67.4	18.2	-
Invert Sugars (g)	30-41	-	-
Ash (g)	2.9	1.5	0.72
Calcium (mg)	35-170	101	35.5
Phosphorus (mg)	54-110	140	45.6
Iron (mg)	1.3-10.9	5.2	1.5
Sodium (mg)	24	-	-
Potassium (mg)	375	-	-
Thiamine	0.16	0.24	0.072
Riboflavin	0.07	0.17	0.148
Niacin	0.6-0.7	4.1	1.14
Ascorbic Acid	0.7-3.0	3.0	13.8
Tartaric Acid (mg)	8-23.8	-	-

Source: Morton (1987)

3.1. CHEMICAL CONSTITUENTS OF *TAMARINDUS INDICA*

Phytochemical study on *T. indica* indicate the presence of constituents, like; phenolic compounds, cardiac glycosides, tartaric acid, the mucilage, pectin, fatty acids and essential elements; As, Ca, Cu, Fe, Mn, Mg. Tamarind pulp typically contains 20.6 per cent water, 3.1 per cent protein, 0.4 per cent fat, 70.8 per cent carbohydrates, 3.0 per cent fibre and 2.1 per cent ash, thus the pulp has a low water content and a high level of protein, carbohydrates and minerals (Table 3.1). The chemical composition of amino acids, fatty acids, and minerals of tamarind plant parts has been widely reported. Large number of investigations and scientific validations has been made to justify a detailed and intricate chemical composition profile of the various parts of the tamarind seed and rest of the parts of the plant. A few such salient findings of reputed scientific investigations are reported below

3.2. FRUITS

Nutritional composition of tamarind fruit varies considerably depending on the climatic conditions of its growth, variations in genetic strains and different stages of maturity. A typical fruit contains 40 per cent pulp. Based on few other reports, the fruit contains about 55 per cent pulp, 34 per cent seeds, and 11 per cent shell (pod) and fibres.

3.3. SEEDS

Fatty acid like palmitic acid, oleic acid, linoleic acid, and eicosanoic acid. The unsaponifiable matter from the seed oil like: β-amyrin, compesterol, β-sitosterol. Polyphenolics was dominated by proanthocyanidins in different forms like apigenin, catechin, procyanidin B2, along with taxifolin, eriodictyol and naringenin. Tamarind seeds comprised only procyanidins, represented mainly by oligomeric procyanidin.

3.4. PULP

Tamarind pulp contains different organic acids like: tartaric acid, acetic acid, citric acid, formic acid, malic acid, and succinic acid. Other classes of compounds reported are the triterpenoids, lupanone and lupeol and some essential oil with benzyl benzoate and limonene as major compounds.

Pulp is a good source of the B vitamins (thiamin, niacin, and riboflavin), Vitamin 4.79 (mg) per 100 g, Vitamin E 93.16 (mg) per 100 g (μg of α-tocopherol) and foltaes 59.35 (μg) per 100 g of pulp; as well as phosphorus, potassium, and calcium (whose content is reportedly the highest found in any fruit)

One of the most versatile characteristics of tamarind namely its sweet acidic taste, is due to mostly tartaric acid (1%). The latter is synthesised in tamarind leaves in the light and translocated to the flowers and fruits. Tartaric is an unusual plant acid formed from the primary carbohydrate products of photosynthesis, and once formed. The content of tartaric acid does not decrease during fruit ripening, suggesting it is not utilized in fruit development. At this same time of fruit development; reducing sugars increase to 30-40 per cent giving the sour fruit a sweeter taste (El-Siddig *et al.*, 2006).

The dry pulp also consists of about 4 per cent protein, and 1 per cent fat.

3.4.1. Amino Acid Profile

The fruit pulp is relatively poor in protein though the fruit is rich in several amino acids (Ishola *et al.*, 1990). Glew and co-workers (2005) reported the amino acid composition of tamarind fruit pulp (Table 3.2).

Table 3.2 Amino acid profile of Tamarind Pulp

Amino Acid	mg/g dw
Crude protein (Total protein)	116.00
Aspartic acid	12.00
Glutamic acid	16.70
Serine	6.88
Glycine	5.15
Histidine	3.37
Arginine	8.74
Threonine	6.05
Alanine	6.20
Proline	7.61
Tyrosine	4.34
Valine	6.97
Methionine	2.48
Isoleucine	5.20
Leucine	8.89
Phenylalanine	4.78
Lysine	8.22
Cysteine	1.35
Tryptophan	1.04

Source: Glew *et al.* (2005)

3.4.2. Fatty Acid Profile

Tamarind fruit pulp is relatively poor in oil (25.3 g/kg of crude lipid), greenish yellow in colour and liquid at room temperature. Saponification values of the oil are high, indicating that it contains a high proportion of low molecular weight fatty acids (Ishola *et al.*, 1990). The fruit pulp contains very little essential fatty acids, linoleic acid and even lower amounts of alpha linolenic acid (Table 3.3).

Table 3.3 Fatty acid profile of Tamarind Pulp

Fatty Acid	μg/g dw
C12:0	0.01
C14:0 Myristic	0.01
C15:0	0.01
C16:0 Palmitic	1.80
C16:1n-7 Palmitoleic	0.12
C18:0 Stearic	0.70
C18:1n-9 Oleic	2.29
C18:1n-7	0.55
C18:2n-6 Linoleic	3.42
C18:3n-3 α-linolenic	0.21
C20:0 Arachidic	0.07

Source: Glew *et al.* (2005)

3.4.3. Mineral Composition

The fruit pulp is a rich source of several macro and micro elements including relatively high amounts of copper, manganese and zinc. Fruit pulp is also a good source of calcium and phosphorus (El-Siddig *et al.*, 1999), but is low in iron (Glew *et al.*, 2005).

3.5. TRADITIONAL USES OF *TAMARINDUS INDICA*

3.5.1. Leaves

Reduce inflammatory swelling, tumours, ring worm, diseases of blood, small pox, xerophthalmia and other eye disease, ear ache, snake bite. Tamarind leaves are dried or boiled are used as poultices for swollen joints, sprains and boils. The leaves, mixed with salt and water, are used to treat throat infections, coughs, fever, intestinal worms, urinary troubles and liver ailments.

3.5.2. Flower

Appetizing, urinary discharges, bad odour in perspiration. They are usually applied after grinding with leaves into powder whereby they are used in lotions or infusions. Lotions and extracts made from them are used in treating conjunctivitis, as antiseptics, as vermifuges, treatments for dysentery, jaundice, erysipelas and haemorrhoids, and various other ailments.

3.5.3. Unripe fruit pulp

Astringent, to the bowel and cure

3.5.4. Ripe fruit pulp

Appetizing, laxative heating, tonic to the heart, anthelmintics and cure heals wound and fracture, biliousness and bile disorder. Pulp act as a cholagogue, laxative and are often used in treating liver 'congestion', constipation and haemorrhoids.

3.5.5. Bark

The bark of the tamarind tree is regarded as an effective tonic and febrifuge. It is used as a tonic and in lotions or poultices to relieve sores, ulcers, boils and rashes. Fried with salt and pulverized to an ash, it is given as a remedy for indigestion and colic. A decoction is used in cases of gingivitis, asthma and eye inflammations. Lotions and poultices made from the bark are applied on open sores and caterpillar rashes. The bark of the tree should be peeled off if needed for medicinal purposes during the time when the tree is not flowering or when the flowering season ends.

3.5.6. Root

Root is used as traditional medicine for the management of diabetes mellitus. The fruit has also been considered as an effective digestive, carminative, laxative, expectorant and blood tonic. Many other properties have been also reported like hypolipemic and antioxidant, anti-inflammatory, antimicrobial, cytotoxic and cure gastrointestinal spasms.

3.5.7. Seed

The powdered seeds are made into a paste for drawing boils and, with or without cumin seeds and palm sugar, are prescribed for chronic diarrhoea and dysentery. The seed coat, too, is astringent, and is also specified for the latter disorders. An infusion of the roots is believed to have curative value in chest complaints and is an ingredient in prescriptions for leprosy.

3.6. PHARMACOLOGICAL ACTIVITY OF *TAMARINDUS INDICA*

3.6.1. Antidiabetic, Hypolipidemic and antioxidant

Hydroalcoholic and aqueous extract of seed of *T. indica* poses significant antidiabetic activities. The results of antioxidant activity clearly exhibit the antioxidant property of ethanolic extract of *T. indica* on Streptozotocin induced diabetic rats. Hydroalcoholic and aqueous extracts of *T. indica* seeds possess hepatoprotective and antioxidant activities. Significantly decreased the hepatic function test markers like SGOT, SGPT, ALP and serum bilirubin and significantly increase the antioxidant enzyme like GSH, CAT, SOD and significantly decreased lipid peroxidation.

3.6.2. Antimicrobial

Ethanolic extract of leaf and stem *T. indica* extracts shows antibacterial activity against some gram negative bacteria. Hydroalcoholic and aqueous extracts of *T. indica* leaves possess antimicrobial activity against some gram positive and negative bacteria like: *S. aureus, B. subtilis, E. coli* and *P. aeruginosa*.

3.6.3. Laxative activity

It improves movement of bowel it is a mild laxative due to the presence of tartaric acid and malic acid in the salt form.

3.7. FOOD USES

3.7.1. Fruits

Tamarind fruit pulp is used for seasoning, as a food component, to flavour confections, curries and sauces, and is a main component in juices and certain beverages.

3.7.2. Seeds

Tamarind seed is a by-product of the tamarind pulp industry. The presence of tannins and other dyeing matter in the testa make the whole seed unsuitable for direct consumption. The major industrial product of tamarind seed is the tamarind kernel powder (TKP) which is an important sizing material used in the textile, paper, and jute industries (Kumar and Bhattacharya, 2008). Tamarind seed is also the raw material used in the manufacture of polysaccharide (Jellose), adhesive and tannin. Flour from the seed may be made into cake and bread. Roasted seeds are claimed to be superior to groundnuts in flavour.

In view of the overall nutrient and chemical composition, tamarind seeds may be adopted as an inexpensive alternative protein source to alleviate protein malnutrition among traditional people living in developing countries (Siddhuraju, 2007).

3.7.3. Fibre

One of the health benefits of tamarind is that the dietary fiber obtained from this herb is beneficial to naturally treat digestive problems such as loose motions and also for the prevention of constipation as the fiber tends to add bulk and which is required for healthy bowel movements. This fiber also helps in binding together any toxins that may be present in the food and thus protects the mucus membrane of the colon from chemicals that cause cancer.

The nutritional value of the tamarind herb is also tremendous as the dietary fiber obtained from tamarind bind the bile salts that are produced from cholesterol and thus minimize their re-absorption into the colon. The tamarind herb is therefore beneficial in lowering the levels of bad cholesterol in the body.

3.8. OTHER HEALTH BENEFITS

The health benefits of tamarind also include it being effective on regulating the blood pressure and the heart rate on account of the presence of potassium in tamarind. Tamarind is also rich in other minerals such as calcium, iron, copper, zinc and magnesium. The iron present in tamarind helps in the production of red blood cells. It is used as a digestive, laxative and also as a natural treatment for bile disorders and biliousness. Therapeutic uses of tamarind also include the juice of the tamarind herb flowers being used in the natural treatment of bleeding piles. The medicinal uses of tamarind also include preparing a mixture of the pulp, flowers and leaves and then applying this on the joints in case of swelling.

In the case of dry eye syndrome, one of the medicinal uses of tamarind is to use eye drops that are prepared from the extracts of the tamarind seeds.

REFERENCES

El-Siddig, K.G. Ebert and P. Ludders (1999). Tamarind (*Tamarindus indica* L.) A review on multipurpose tree with promising future in Sudan. *Journal of applied Botany – Angewandte Botanik,* **73**: 202-205.

El-Siddig, Gunasena, Prasad, Pushpakumara, Ramana, Vijayanand and Williams (2006). Tamarind (*Tamarindus indica*). Southampton Centre for Underutilised Crops, Southampton, UK

Glew, R.S., Vander Jagt, D.J., Chuang, L.T., Huang, Y.S., Millson, M. and Glew, R.H. (2005). Nutrient content of four edible wild plants from West Africa. *Plant Foods for Human Nutrition,* **60**: 187-193.

Ishola, M.M., Agbaji, E.B. and Agbaji, A.S. (1990). *A Chemical Study of Tamarindus indica (Tsamiya) Fruits Grown in Nigeria. Journal of the Science of Food and Agriculture,* **51:** 141-143.

Kumar, C.S. and S. Bhattacharya (2008). Tamarind Seed: Properties, Processing and Utilization. *Critical Reviews in Food Science and Nutrition,* **48**: 1-20.

Morton, J. (1987). Tamarind, *In:* Fruits of Warm Climates, Miami FL, Pp: 115-121.

Siddhuraju, P. (2007). Antioxidant activity of polyphenolic compounds extracted from defatted raw and dry heated *Tamarindus indica* seed coat, 40: 982–990.

■■■

Chapter **4**

HARVEST AND POSTHARVEST PROCESSING OF TAMARIND

Dr. N. Karpoora Sundara Pandian* & Dr. P. Rajkumar**

**Assistant Professor, College of Food and Dairy Technology, Tamil Nadu Veterinary and Animal Sciences University, Chennai – 600 052*
***Professor and Head, Department of Agricultural Processing and Basic Sciences, Agricultural Engineering College and Research Institute, Kumulur, Trichy – 621 712*

Tamarind is a very common souring agent in all Indian cuisine due to the presence of tartaric acid, biochemically dihydroxydicarboxylic acid, $C_4H_6O_6$. Tamarind is harvested during January to April of every year near to summer season in all over India at the moisture content of 25 per cent - 30 per cent on dry basis. The hull at unripe tamarind sticks intact with pulp. As ripening process advances, the pulp and hull is separated as consequences of reduction in pectin content. The hull is green and rigid in nature at the immature stage. The brown colour and brittleness of hull are the maturity indices of tamarind.

On an average, the whole tamarind fruit contains 55 per cent pulp, 34 per cent seeds and 11 per cent hull and fibres (Rao and Mathew, 2001). The whole tamarind fruits are oblong or sausage shaped, curved or straight with rounded ends. The hull is brittle in nature and it is covered on the pulp. The pulp is analogous to human finger in shape connected by a series of compartments and each compartment is possessed of a single seed. The outer surface of the pulp has three to four tough branched fibres from the base to the apex to strengthen the compartments. The pulp is sticky in nature as it is highly hygroscopic. The tamarind pulp is rich in sugar ranging from 21 per cent to 30 per cent and the hygroscopicity of the pulp increases with increase in relative humidity while storing under atmospheric temperature. The stickiness of the tamarind is experienced at the moisture content of 25 per cent on dry basis (Pandian, 2015).

4.1. HARVEST

Generally, the tamarind fruits are left to ripen in the tree itself and after reaching mature stage, harvesting operation is performed. Conventionally, the sour tamarind ripe fruits are harvested by shaking the branches and the whole tamarind fruits are collected on a spread mat beneath the tree whereas the sweet tamarind fruits in Thailand are handpicked and sold in the market. The lengthy wooden bamboo poles are sometimes used in picking of whole tamarind fruits (Gunasena and Hughes, 2000).

Trunk shaker and branch shakers are the few mechanical advancements in harvesting of fruit bunches. Trunk shakers are best suited for trees that have soft trunks and branch shakers are best suited for trees that have hard trunks like tamarind. Nowadays, branch shakers are available at 2.5 kW power level. The shaker is powered by two stroke petrol engine. This commercial branch shaker is similar to brush cutter with 2 m long hooked pipe at the end to hold the branches intact and reciprocal action of the engine imitates the shaking action (Anonymous, 2015).

4.1.a. Physical and Frictional Properties of Tamarind

The knowledge of physical and mechanical properties helps to analyze the behaviour of tamarind during handling and designing of process equipment. The properties such as moisture content, size, shape, bulk density, true density, porosity, angle of repose, surface area and coefficient of friction were determined for whole and dehulled tamarind purchased from Hosur, Tamil Nadu (Pandian *et al.,* 2013 and Pandian, 2015).

Average values of length, width and thickness of the whole tamarind fruits at 22 per cent moisture content on dry basis were 93.87 mm, 21.11 mm and 14.90 mm, respectively. Geometric mean diameter, sphericity index and surface area were found to be 30.91 mm^2, 0.33 mm^2 and 3000.18 mm^2 respectively. The properties such as bulk density, true density, porosity and angle of repose whose average values were 240.39 kg/m^3, 469.59 kg/m^3, 47.80 per cent and 35.4°C respectively. Coefficient of friction on different surfaces namely, wood, mild steel, stainless steel, aluminium and galvanized iron sheet were found for whole tamarind fruit, whose average values were 0.45, 0.58, 0.33, 0.40 and 0.40 respectively (Pandian, 2015).

The average length, width, thickness, geometric mean diameter, sphericity index, surface area, bulk density, true density and porosity of dehulled tamarind fruit at the moisture content of 25 per cent on dry basis were determined. Angle of repose during filling and emptying were in an average of 33.53°C and 49.72°C respectively. Coefficient of friction for tamarind fruits on different surfaces namely, Hard board, mild steel, stainless steel, aluminium and galvanized iron sheet were found to be 0.43, 0.59, 0.41, 0.48 and 0.58 respectively (Pandian *et al.,* 2013).

4.2. POST HARVEST OPERATIONS

4.2.1. Traditional Method of Tamarind Processing

Generally, wet and dry methods are followed in tamarind processing. Tamarind is traditionally processed in dry way which commonly includes the unit operations of drying of whole tamarind to keep away sticking of pulp with hull, dehulling, defibring, deseeding, pressing into cakes and storage. The dehulling operation of tamarind involves sun drying of harvested whole tamarind and then hull is removed from the pulp by beating with sticks (Fig. 4.1). Seed removal is one of the important and labour intensive unit operations in tamarind processing and conventionally different methods are followed.

Fig. 4.1. Labors beating the dried whole tamarind as part of dehulling

Fig. 4.2. Women workers are deseeding tamarind at Krishnagiri, Tamil Nadu

The current practice of tamarind deseeding in northern part of Tamil Nadu is to impact the vertically oriented fruits manually with a hammer or wooden mallet by women labourers (Fig. 4.2). In the middle part of Tamil Nadu, tamarind is deseeded by the process of hand pounding in which a stone mortar is sprayed with oil, generally castor oil, and a wooden pestle is used to exert impact load over the fruits. In Southern parts of Tamil Nadu, a knife or long sharp needle is used to deseed. The current conventional methods (Fig. 4.3 and 4.4) followed are crude, unhygienic, labour intensive and time consuming. Hence, in order to facilitate tamarind processing, machineries were developed and released from state Agricultural Universities for the benefit of the processors and to ensure hygienic practices of processing.

Fig. 4.3. Defibering action

Fig. 4.4. Pressing action of deseeded tamarind into rubber ring to form cake

4.2.2. Drying of whole and dehulled tamarind

Dehulling and deseeding are the major unit operations in tamarind processing before which tamarind fruits are subjected to drying to mitigate the stickiness of pulp with physical and mechanical parts.

A local variety of whole and dehulled tamarind purchased from Hosur, Tamil Nadu was dried under open sun (Fig. 4.5) and tray dryer at temperatures of 50, 60 and 70°C. It was observed that the mechanical drying of tamarind at 70°C had higher moisture removal rate followed by drying at 60°C and 50°C. Drying of tamarind occurred in the falling rate period and constant rate period was not found (Pandian and Rajkumar, 2015). The drying data for sun drying of whole tamarind was well fitted with Midilli *et al.* Model. The drying kinetics at 50, 60 and 70°C followed the Jena-Das model, Midilli *et al.* model and Verma *et al.* model, respectively (Pandian, 2015).

Fig. 4.5. Sun drying of whole tamarind

4.3. MECHANIZATION

4.3.1. Machineries for Dehulling

A tamarind dehuller developed at University of Agricultural Sciences, Bengaluru, India has a capacity of 500 kg/h. The dehulling efficiencies are reported as 80 per cent and 58 per cent for the large size curved fruits and small fruits, respectively. The tamarind dehulling unit consists of a mild steel frame, a hopper, shafts, dehulling rings and a 1 hp motor. The shafts mounted with dehulling rings are made to rotate in opposite direction by means of a V groove pulley. The whole tamarinds after drying are fed into the clearance between the dehulling rings. The pulp and husk separated due to impact force.

A dehuller for tamarind developed at Tamil Nadu Agricultural University (TNAU), Coimbatore, India has a capacity of 100 kg/h with the dehulling efficiency of 94 per cent. The mechanism involves sharp 'L' shaped pegs made up of 15 x 3 mm size mild steel flat mounted on the central shaft that encased with 20 x 5 mm mild steel oblong sieve. The principle of the machine (Fig. 4.6) is application of impact and simultaneous shear.

Fig. 4.6. Tamarind Dehuller available at Tamil Nadu Agricultural University, Coimbatore

4.3.2. Machines for Deseeding

A tamarind seed expeller designed and developed at University of Agricultural Sciences, Bengaluru is reported to contain, the pulp-squeezing unit with serrated roller mounted fitted on a mild steel frame and a mild steel rasp bar. The speed of the roller is maintained at 300 rpm. The roller is attached to a one hp motor by means of a V belt pulley arrangement. The dehulled and dried tamarinds are fed into the clearance between roller and rasp bar. The results of the trials conducted indicate that the mechanical expeller is nearly 11 times faster than the traditional seed expulsion and has a capacity of 35 kg/h at 22% pulp moisture content on wet basis (Babu *et al.*, 1999).

A tamarind deseeder developed at Tamil Nadu Agricultural University with the principle of impact and simultaneous shear (Fig. 4.7) is used widely in Krishnagiri and Dharmapuri districts of Tamil Nadu to deseed small sized tamarinds. The deseeding efficiency of the machine is 83% and cost is Rs. 20,000/- (Pandian, 2010).

Fig. 4.7. Tamarind deseeder available at Tamil Nadu Agricultural University, Coimbatore

A hammer type tamarind deseeder developed and evaluated at Kumulur, Tamil Nadu Agricultural University consists of a feeding roller with rubber lining, hammering mechanism, motor and power transmission system and frame to support the mechanism. The machine operate on the principle of impact and, deseeding efficiency of the machine is found to be 79 per cent at 5 rpm (0.06 m/s); peripheral speed of feeding roller to deseed the tamarind at the moisture content of 22.5 per cent on dry basis providing lowest mechanical damage to seed (0.3%) and pulp (14.94%). The mechanism (Fig. 4.8) is found suitable for the large sized fruits containing more than three seeds. Feeding of small sized fruits such as single and double seeded fruits in the feeding roller is difficult due to the variation in shape (Pandian and Rajkumar, 2014).

A tamarind deseeder is developed and released during 2016 at Tamil Nadu Agricultural University to get full strip of tamarind in halves after deseeding. The mechanism is found suitable to all sized tamarind fruits. The deseeding efficiency of the developed machine is 92 per cent and seed damage is found to be 8.5 per cent.

Fig. 4.8. A hammer type tamarind deseeder available at Tamil Nadu Agricultural University, Kumulur, Trichy

4.3.3. Wet Method of Tamarind Pulp Extraction

For the extraction of tamarind pulp, the whole dehulled tamarind fruit is soaked in hot water or cold water. The wet pulp is passed through a metallic screen to get a pure pulp which is free from fibre and seed. Then the extracted pulp (slurry) is either concentrated or dehydrated into powder (Benero *et al.*, 1972 and Baragano de Mosqueda, 1980).

Deseeded tamarind pulp is the fast moving value added product in the domestic market whereas the tamarind concentrate and tamarind pulp powder are the exportable value added products from tamarind.

4.4. STORAGE OF TAMARIND

Tamarind fruit after processing in dry method is conventionally stored in mud pot at kitchen where dry weather exists in order to avoid absorption of moisture from the surroundings. In Thailand, sour tamarind fruit after deseeding are mixed with salt and balled to store in poly packs for retail marketing. In Tamil Nadu, dry processed tamarind fruit is pressed inside out into cake form to avoid moisture absorption of moisture and covered with plastic packs for retail marketing.

Commercially, dehulled tamarind is stored at cold storage units below 10°C in order to prevent enzymatic browning. Brownish red of tamarind is the acceptable colour in the market and treated as fresh from farm. Dark brown or black tamarind is treated as old stock and hence, the dark tamarind is sold at throwaway prices. The enzyme, Polyphenoloxidase act on the pulp in the presence of oxygen soon after the break of hull. The peroxidase and non-enzymatic browning increase with storage period (Kotecha and Kadam, 2003). The low temperature storage is recommended and adopted technology to mitigate the enzyme activity in order to prevent the colour change.

Generally, the maximum shelf life of tamarind is one year under dry weather whereas prolonged storage under wet condition, the pulp becomes soft due to moisture absorption and pectolytic degradation takes place. Hence, proper storage method and selection of packaging material is important in prevention of colour change and moisture migration.

Cold storage of pulp in transparent containers at 0°C and pulp mixed with powdered salt at 10% in a transparent container are found to be satisfactory ways to prevent the colour change in tamarind (Feungchan *et al.*, 1966).

Chumsai-Silvanich *et al.* (1991) recommended steaming and immediate drying of the tamarind fruit at 80°C for two hours and 60°C for 2.5 hours in a hot air oven and packing in clear plastic bags for extending the shelf life of tamarind for three months.

Deseeded tamarind pulp packed in 150 gauge poly bags stored under 4°C for up to six months retain reddish brown colour (Puranaik *et al*, 2004). Similar results were found by Nagalakshmi and Chezhian (2004). Vacuum packed deseeded tamarind pulp in 800 gauge poly bags stored up to 330 days under refrigeration at 4 ± 20°C exhibit no colour change in the pulp right from the initial stage of storage.

The overall acceptability of tamarind stored at 27°C and 65% RH for 180 days are found to be good for the tamarind pulp packed in aluminum foil under vacuum, aluminum foil, metallized polyester, polyethylene and polypropylene pouches. A study indicated that the pulp loss percentage during storage is low in black polyethene and plastic compared to phoenix mat and metal (Ramkumar *et al*, 1997).

4.5. FUTURE SCOPE FOR TAMARIND PROCESSING

1. Mechanized processing of tamarind without human intervention is expected due to raising awareness on Good Manufacturing Practices (GMP) and huge labour shortage.

2. Low cost technologies should be developed to mitigate the colour change of tamarind pulp during storage.
3. Exclusive Dryers and Material Handling Equipments are needed to be developed towards complete mechanization.
4. Storage studies to prevent insect infestation, fungal spoilage and enzymatic reaction need to be developed with the aid of Modern Science and Technology.
5. More value added products are needed to be developed.
6. Development and popularization of traditional food products from tamarind using modern techniques is expected to empower the farming community.

REFERENCES

Anonymous (2015). Web source. http://www.aliexpress.com/item/Olive-harvester machine-nut-olive-shaker-tree-shaking-machine-tree-branch-shaker vibration - machine -fruit-branch/32319339259.html. Dt: 12.06.2015.

Babu, C. K., M. V. Ramkumar and B. Ranganna (1999). Post harvest management of tamarind. All India Coordinated Research Programme Manual. pp: 1-54.

Baragano de Mosqueda, M. (1980). Technology of clarified tamarind juice. In tropical and sub tropical fruits. *AVI Publishing.* Westport, pp: 375.

Benero, J.R., A.J. Rodriguez and Collazo de Rivera (1972). Tamarind. Journal of the Agricultural University. *Puerto Rico.* **56(2)**: 185-186.

Feungchan, S., T. Yimsawat, S. Chindaprasert and P. Kitpowsong (1996). Studies on the colour change of tamarind pulp in storage. *Thai Journal of Agricultural Science,* **(1)**: 52-56.

Kotecha, P.M. and S.S. Kadam (2003). Studies on browning in tamarind pulp during storage. *Journal of Food Science and Technology.* **40(4)**: 398-399.

Nagalakshmi, S. and Chezhian, N. (2004). Influence of type of packaging and storage conditions on Tamarind (*Tamarindus indica*) pulp quality. *Journal of Food Science and Technology,* **41(5)**: 586-590.

Pandian, N.K.S. (2010). Development and Evaluation of Tamarind Seed Remover. Unpublished *M.Tech. thesis.* Department of Food & Agricultural Process Engineering. Tamil Nadu Agricultural University. Coimbatore. India.

Pandian, N.K.S., P. Dhananchezhiyan and S. Parveen (2013). Physical and Engineering Properties of Tamarind Fruit. *International journal of Scientific Engineering and Technology,* **2(11)**: 1083-1087.

Pandian, N.K.S. and P. Rajkumar (2014). Development and Evaluation of Hammer type Tamarind (*Tamarindus indica* L.) deseeder. *Research Journal of Agricultural Sciences,* **5(6):** 1228-1231.

Pandian, N.K.S. (2015). Development and Evaluation of Tamarind (*Tamarindus indica* L.) Dehuller and Deseeder. Unpublished *Ph.D. thesis.* Department of Food &

Agricultural Process Engineering. Tamil Nadu Agricultural University. Coimbatore. India.

Pandian, N.K.S. and P. Rajkumar (2015). Sun and Mechanical drying and study on drying rate kinetics of tamarind (*Tamarindus indica* L.) at different temperatures. *Environment & Ecology,* **34(1A)**: 324-328.

Ramkumar, M.V., C.K. Babu, S. Subramanya, B. Ranganna and K.C. Krishnamurthy, (1997) Development of a tamarind dehuller and short-term storage of pulp. Proceedings of National Symposium on *Tamarindus indica* L., Tirupathi (A.P.) Organised by the Forest Department of A.P. 27-28 June 1997: 145-150.

Rao, Y.S. and K.M. Mathew. (2001). Handbook of Herbs & Spices. Eds. K.V. Peter. Wood head publishing limited. England. Pp: 102.

■■■

Chapter **5**

VALUE ADDITION AND BY PRODUCTS

Dr. Reni

Assistant Professor, Department of Food Processing and Preservation Technology Avinashilingam University

5.1. TAMARIND PULP

The most valuable and commonly used part of the tamarind tree is the fruit pulp. The pulp constitutes 30-50 per cent of the ripe fruit (Purseglove, 1987), the shell and fibre account for 11-30 per cent and the seed about 25-40 per cent (Chapman, 1984; Shankaracharya, 1998). Pulp is rich in pectin and reducing sugars and contains significant amounts of organic acids, 98 per cent of which is tartaric acid. The main flavour compound of the pulp is 2-acetylfuran.The tamarind tree bears pods (fruits) containing about 10 brown seeds surrounded by an abundant acid pulp. The fruits are harvested when ripe; pulp is rust-coloured and contains 38 per cent moisture content. The most outstanding characteristic of tamarind is its sweet acidic taste, the acid due to mostly tartaric acid (10 %). Tartaric is an unusual plant acid formed from the primary carbohydrate products of photosynthesis, and once formed, it is not metabolically used by the plant. The content of tartaric acid does not decrease during fruit ripening, suggesting it is not utilized in fruit development. At this same time of fruit development; reducing sugars increase to 30-40 per cent giving the sour fruit a sweeter taste. In intertropic zone, tamarind pulp leaves and flowers are commonly used in various dishes or traditional drinks due its high nutritive and calorific value. The Nutritive value per 100 gram of Fruit Pulp, Leaves and flowers are given in Table 5.1.

5.1.1 Harvesting of Tamarind Fruit Pulp

Harvesting of tamarind fruits commences when the pods feel brittle by pressing. Trees grown from may take more than seven years to start bearing and up to 10 or 12 years. During maturity the Tamarind seed pods fill (Fig. 5.1). The pulp becomes brown to reddish brown and the testa becomes brittle and cracks easily. The pulp dehydrates and becomes sticky and the seeds become hard, dark brown and glossy mature fruits have brown shells, while immature ones are green.

Tapping them with the finger can help to identify the mature fruits. A hollow, loose sound will be produced as the pulp shrinks with maturity and the shell becomes brittle. At this stage the fruits are mature and ready for harvesting. However, it is not always easy to determine whether the fruits are ready for harvesting, as the testa colour only changes slowly as the pods mature. Individual fruits on the same tree also mature at different times making it necessary for harvesting to be done selectively.

Table 5.1 Nutritive value per 100 g of edible Portion

Constituents	Pulp (Ripe)	Leaves (young)	Flowers
Calories	115	-	-
Moisture (g)	28.2-52.0	70.5	80.0
Protein (g)	3.10	5.8	0.45
Fat (g)	0.1	2.1	1.54
Fibre (g)	5.6	1.9	1.5
Carbohydrates (g)	67.4	18.2	-
Invert sugars (g)	30-41	-	-
Ash(g)	2.9	1.5	0.72
Calcium (mg)	35-170	101	35.5
Phosphorous (mg)	54-110	140	45.6
Iron (mg)	1.3-10.9	5.2	1.5
Sodium (mg)	24	-	-
Potassium (mg)	375	-	-
Thiamine (mg)	0.16	0.24	0.072
Riboflavin (mg)	0.07	0.17	0.148
Niacin (mg)	0.6-0.7	4.1	1.14
Ascorbic acid (mg)	0.7-3.0	3.0	13.8
Tartaric acid (mg)	8-23.8	-	-

Source: Morton (1987)

In the Philippines fruits are harvested at two stages, green for flavouring and ripe for processing. The fruits of sweet types are also harvested at two stages, half ripe and ripe. At the half ripe stage the pulp is yellowish green and has the consistency of an apple, particularly in the case of sweet forms. At the ripe stage, the pulp shrinks, due to loss of moisture and changes to reddish brown and becomes sticky. Tamarind is not generally a dessert fruit although the sweet tamarind is often eaten fresh directly from the pod. In most countries the sour tamarind ripe fruits are harvested by shaking the branches and the pods are collected on a mat. Sweet tamarind pods fetch a high price both in local and foreign markets and are used to pick the fruits. If the whole fruit is marketed, both sweet and sour types should be harvested by clipping to avoid damaging the pods. The use of poles in picking is not desirable as knocking can cause damage to the pods (Fig. 5.1).

Fig. 5.1. Tamarind Fruit Pulp

In India the fruits are left to ripen on the tree before harvesting, so that the moisture content is reduced to about 20 per cent. When the fruits are left un-harvested they may remain hanging on the tree for almost one year after flowering and eventually abscise naturally. Sometimes they remain on the tree until the next flowering period. Tamarind fruit, as a non-climacteric, will not ripen any further after harvest.

5.1.2. General Processing of Tamarind Fruit Pulp

Fresh fruits are often dried using small-scale dehydrators, however in most countries rural households dry pods in the sun, the shells, fibres and seeds are then removed and the pulp stored in plastic bags or earthenware pods.

Processing consists of collecting fresh fruits, drying either in sun or dryers, cracking fruits and separating pulp and fiber's from broken shells, cleaning to remove seeds and packing of dry pulp either in polythene covers or air tight PET bottles. In most of the tamarind growing countries the dry ripe pods can easily cracked and the pulp and the fibres separated from the broken shells. The pulp is then processed by peeling and removing the fibre strands from the pulp. After separating the pulp from the fibres, seeds and shells, it is compressed and preserved in large masses and solid in small shops and marketed by weight. Sometimes the cleaned pulp is compressed and packed in palm leaf mats or gunny bags for storage and marketing. In east India, the pulp is covered with salt, rolled into balls, exposed to dew and stored in earth ware jars (Chapman, 1984; Morton, 1987), whereas in java, the salted pulp is rolled into balls, steamed and sun dried, then exposed to dew for a week before packing in stone jars. When tamarind is processed for shipment to large scale processors, the pulp is removed from the seeds, layered with sugar and then covered with boiling sugar syrup.

In Sri Lanka, the harvested pods are dried in the sun for 5-7 days to bring all fruits including the half-mature fruit to the fully ripe stage. The pods are then cracked to separate the pulp. The pulp with the seed is spread on thick polythene sheets and dried in the sun for 3-4 days to remove excess moisture and prevent the growth of moulds. The dried pulp is mixed with salt and packed in clay pots for storage. These pots are usually kept in a dry place. Through this process tamarind can be stored for about a year, after which the colour changes to dark brown or black and changes in flavour occur. Although pulp is stored with the seeds intact in rural houses holds, seedless pulp is stored in plastic bags in retail shops (Guna Sena, 1997).

5.1.3. Storage of Tamarind Fruit Pulp

The high soluble solids content and the low water content of tamarind fruit contributes to its long storage-life tamarind can be stored with the shell or as a separated dry pulp and tightly packaged pods can be stored at 20°C for several weeks. The pulp of mature tamarind is commonly compressed and packed in palm leaf mats or plastic bags and stored at 20°C for a significant period when processed into paste. It can be frozen and stored for one year, or refrigerated for up to six months. Under dry conditions the pulp remains good for about one year, after which it becomes almost black. In humid weather, especially, the pulp becomes soft and sticky as pectolytic degradation takes place storage, the dry, dark-brown pulp becomes soft, sticky as pectolytic degradation takes place and moisture is absorbed (Lewis and Neelakantam, 1964) during storage, the dry, dark-brown pulp becomes soft, sticky, and almost black. The pulp can be stored for a longer period after drying or steaming. According to the research finding of CFTRI (Central Food Technological Research Institute, Mysore, India) the pulp could be preserved well for 6-8 months.

Refrigerated storage of the freshly harvested, deseeded, semidried ripe pulp is highly recommended in order to retain the attractive brownish red colour of the pulp for long term storage and marketing. Normally deseeded pulp packed in 150 gauge poly bags of 5 kg capacity (fresh commercial sample) can be stored under refrigerated conditions (4-6°C; RH 75-80%) for up to six months and retain an acceptable brown colour (Puranaik *et al.*, 2004). Recent studies carried out at Tamil Nadu agricultural university, Coimbatore, India, indicated that the freshly harvested deseeded tamarind pulp can be stored for up to 330 days under refrigeration at 4 ± 2°C when vacuum packed in 800 gauge poly bags without any colour change in the pulp right from the initial stage of storage (Nagalakshmi and Chezhian 2004).

Semi-dried tamarind pulp has a moisture content ranging from 17-20 per cent. Tamarind pulp packed in aluminium foil with vacuum, aluminium foil, metallized polyester, polyethylene and polypropylene pouches and stored at 27°C and 65 per cent RH for 180 days retained overall quality.

5.2. PROCESSED PRODUCTS FROM TAMARIND PULP

5.2.1. Tamarind beverages

Tamarind fruit pulp is used for the processing and preparation of beverages. By using Tamarind Pulp good quality ready to serve beverage, syrup and concentrate can be prepared with a shelf life of six months at ambient storage.

5.2.1.1. *Ready to Drink Tamarind beverage*

On a small scale, the fruit pulp is made into a refreshing drink after dissolving in water and squeezing by hand. After removing the extraneous matter, water is added to dilute the drink as required. Tamarind pulp is enjoyed as a refreshing drink and beverage in most of the southern and Central American and Asian countries.

The extraction and processing techniques of the pulp for the preparation of clarified tamarind juice is one in which the structural and colloidal phases are broken to prevent sedimentation which can be done by adding gelatine (Baragano de Mosqueda, 1980). In this process 0.12-0.15 per cent gelatine is agitated with the pulp left to stand for 10-15 days at 10°C. The colloidal particles will precipitate during this period and are removed by decantation process. The obtained juice maintains good colour and flavour. The tamarind drink contained 9-12 per cent pulp and adjusted to 21.5°C Brix. The blended Drink had 10-14 per cent pulp and 15°C and 17°C Brix. Normally the drink after pasteurisation at 185-190°F was canned. The tamarind drink remained in good condition after one year and the blended drink for 10 months when stored at 29.4°C.

Tamarind drinks in poly packs are commercially available in Thailand, Indonesia, and many countries of Africa. In Mexico and Guatemala, the drink is carbonated. It has a high content of Vitamin D (Thiamin and Niacin) as well as small amount of Vitamin C (Ascorbic acid) and Carotene. Tamarind fruit juice is most valued for its ascorbic acid content, which is about 5.12-6.1mg/100ml.

5.2.1.2. *Tamarind Syrup*

Tamarind syrup is made by boiling immature fruit pulp until it is become so soft and strained through cheese cloth. The pulping operation consists of breaking the shells by hand and agitation in water to separate the pulp and the seeds. To one cup of juice, a half-teaspoon of baking soda is added. The mixture is boiled down to one-half of the original quantity, removing the rising scum in the process. The juice is again strained. For very cup obtained, a quarter cup of sugar is added. The mixture is boiled again for 20 minutes. The cooled syrup is poured into sterilised bottles and sealed. The recommended concentration of tamarind pulp in syrup is between 20-24 per cent, so as to produce a beverage with distinctive flavour and acidity. Such syrup contains 56.7 total solids, 43.8 per cent reducing sugars and a total acidity of 1.11 per cent as tartaric acid.

5.2.1.3. *Tamarind juice Concentrate*

Tamarind concentrate is a concentrate of juicy acid pulp prepared from tamarind pulp free from skin, seed, fibers etc. Concentrate can be used for making household preparations like sambars, rasams, chutneys, sauces in south and north Indian dishes. The industrial method used in India for manufacturing tamarind juice concentrate is by extraction of all the soluble in boiling water. Process consists of collecting pulp from tamarind pods and extracting juice from pulp. The pulping operation consists of breaking the shells by hand and agitation in water to separate the pulp and the seeds. The recommended concentration of tamarind pulp in syrup is between 20-24 per cent so as to produce a beverage with distinctive flavour and acidity (Bueso, 1980). In clarified tamarind juice structural and colloidal phases are completely removed by adding gelatine. The juice maintains good colour and flavour and is transparent. Insoluble solids removed and soluble extracts are concentrated under vacuum (Fig. 5.2).

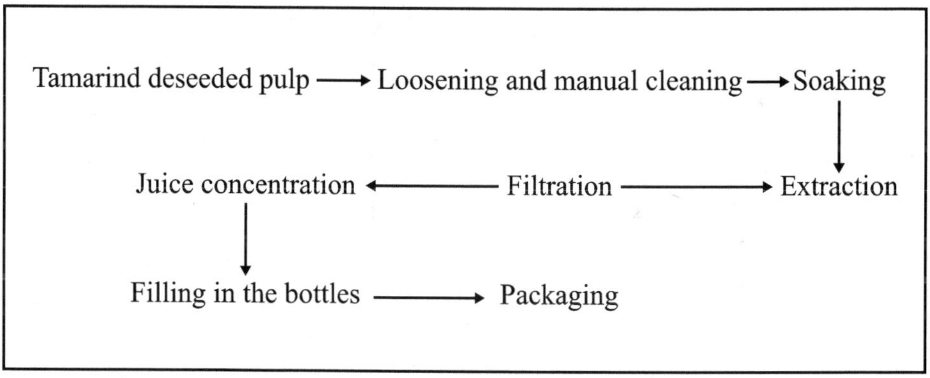

Fig. 5.2. Processing of Tamarind juice Concentrate

A process for the preparation of tamarind juice concentrate was developed by CFTRI, Mysore, India. The process involves extraction of tamarind juice with boiling water using the counter current principle, where dilute extracts are used for extracting fresh batches of the pulp. Using this process, an extract is obtained containing 20 per cent soluble reach 68 per cent, the pulp is placed in cans or bottles. It sets like jam on cooling. The yield of the concentrate is about 75 per cent of the pup used. The quality of the product is assured by using recently harvested fruits free from insect and rodent damage developed to maintain the quality of the product and concentrating the juice in a vacuum evaporator to total soluble solids of 70°Brix. The tamarind concentrate is shelf stable at room temperature and can be diluted and used in various food preparations. It can also be stored for long periods. Tamarind juice concentrate exhibited Newtonian flow behaviour up to 19°Brix and pseudo plastic nature at higher than 23°Brix obeying the power law relationship. The flow behaviour index for concentrate of 23°Brix was 0.747 and 0.625 for concentrate of 28-62°Brix. The chemical composition of tamarind juice concentrate in percentage is given in Table 5.2.

Tamarind juice concentrate gives a sour tinge to the product and enhances flavour of those dishes. It has high content of vitamin B (Thiamine and Niacin) as well as small amounts of vitamin C and carotene16. It is also used in some medicines and for preserving fish. As per international standards the product should have total soluble solids of 45-47 brix with an acid content of 9.0 ± 2.0 per cent coupled with characteristic tamarind flavour and taste.

5.2.2. Tamarind Pulp powder

Tamarind pulp powder (TPP) is one of the convenience food product prepared by concentrating, drying and milling the pulp into a powder form. Technology for processing of raw tamarind pulp into dry powder form was developed by CFTRI. Tamarind Pulp Powder (TPP) is prepared by concentrating, drying and milling the pulp into a powder form.

Depending on the manufacturing process, wide variations in the physico-chemical characteristics are occurred. Tamarind powder can be used as a condiment/adjunct and souring agent in many Indian dishes and sauce. As a convenient product, it has good market potential. Tamarind fruit obtained under standardized conditions of temperature, humidity and mill setting. The equipments needed in manufacture of this product are roaster, planetary mixer, disintegrator, tray drier, weighing machine and Misc. equipments. They reported that average total solid content of tamarind pulp powder varied from 18.6-25 per cent, acidity (as tartaric acid) from 8.7-11.1 per cent with an average value of 9.9 per cent. The moisture content ranges from 3.5-8.8 per cent. Among the minerals; the calcium and potassium contents were high and ranged from 74-143 mg and 23.8-27.7 mg. Starch is the major constituent in tamarind pulp powder (20-41%). The chemical Composition of tamarind Pulp powder in Percentage.

5.2.2.1. *Foam mat dried Tamarind pulp Powder*

Foam-mat drying is a process in which liquid foods are whipped into stable foams and then air-dried. The process consists essentially of the formation of stable foam containing the product to be dried and the air drying of the foam to form a thin porous sheet or mat. The main advantages of foam-mat drying techniques are lower drying temperatures and shorter drying times. These advantages can be attributed to the larger surface area exposed to the drying air which accelerates the moisture removal process. In the Fig. 5.3. Foam mat dried Tamarind pulp Powder is given. However, the lack of foam stability during the heating cycle should be considered. When the foam does not remain stable, cellular breakdown occur causing serious impairment during drying Operation. This disadvantage can be solved by incorporation of the film stabilizer such as polymeric material.

Drying of tamarind pulp by foam mat drying using different foaming agents was reported by Vernon-Carter *et al.* (2001). The foaming agent viz., mequite gum, ovalbum in and a low molecular weight surface active blend were hydrated to 50 per cent (w/w) solutions and added to the samples either singly or in combination. It was found that the mesquite gum with surface active blend and ovalbumin with surface active blend, ovalbum in with mesquite gum resulted in acceptable flavours.

Fig. 5.3. Foam mat dried Tamarind pulp Powder

5.2.2.2. *Drum drying of Tamarind Pulp*

The sour tamarind (*Tamarindus indica* L.) flesh was obtained at the local market. The feed materials prepared and are dried in a double drum dryer (drum diameter 22 cm and length 46 cm) with a 0.5 HP motor using a drum speed of 0.35 rpm and a gap between the drums of 0.4 mm. In the Fig. 5.5, the Schematic diagram of Drum Dryer is given. During the drying experiments, steam pressure is used at the inlet of drum dryer and inside the drum, the weight and the temperature of steam condensate.

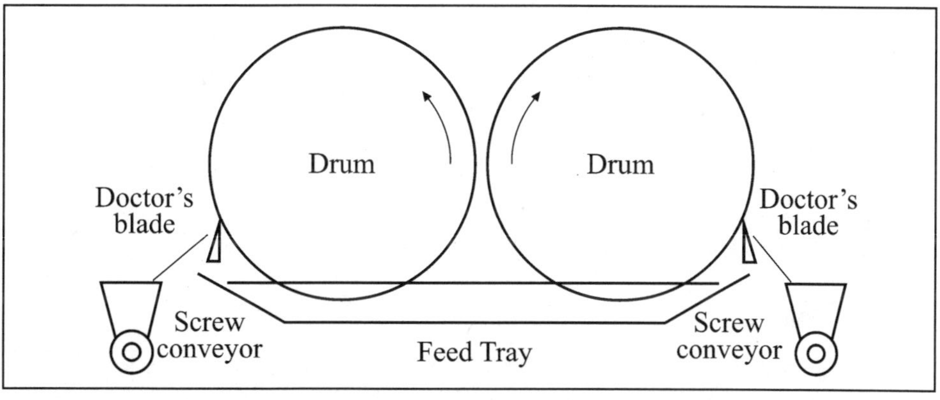

Fig. 5.4. Schematic diagram of Drum Dryer

The quantity of condensate was recorded for the calculation of heat energy consumption. At the end of drying, the tamarind powders were collected, weighed, and kept in the sealed container for the quality determination. The Processing flow chart for Drum drying of Tamarind Pulp is given in Fig. 5.2.

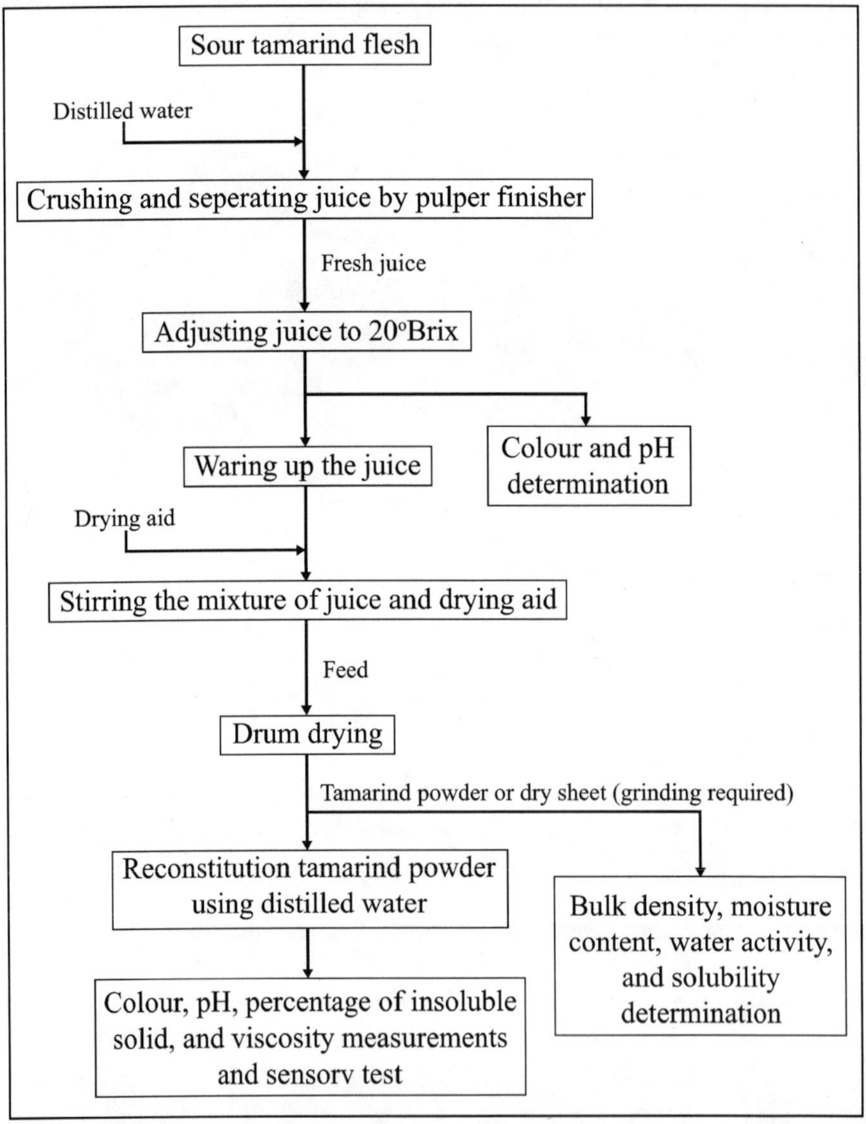

Fig. 5.5. The Processing flow chart for Drum drying of Tamarind Pulp

5.2.2.3. *Fruit Leather*

Fruit leathers prepared from the dried sheets of tamarind fruit pulp will have a soft, rubbery texture and a sweet taste (Fig. 5.6). It can be prepared by drying thin layers of pureed fruit in the dehydrator. It is also called fruit rolls or toffies. This fruit leathers is delicious, wholesome and nutritious high-energy snacks for backpackers, campers and active children. Fruit leather is relatively light in weight, easy to prepare and a good way to use left-over canned fruit and slightly over-ripe fresh fruit. These fruit leathers can be used in pie fillings, in cooking and as a dessert topping.

Fig. 5.6. Tamarind Fruit Leather

5.3.1. Preparation of fruit leather

5.3.1.1. *Extraction of pulp from the fruits*

For the processing of Tamarind fruit leather ripe fruit pods of uniform maturity have to be selected. Then the pulp from the pod has to be separated manually from shell, fiber, rags and seeds.

Cold extraction

The pulp was soaked in the water during cold extraction method. Different Pulp and water ratio was maintained to obtain maximum separation of pulp. The best Pulp: water ratio for extraction is 1:2 for 6 hrs. After soaking the mixture was homogenized and filtered to obtain fine pulp. This extraction rate proportions gives highest total soluble solids and total sugars. The yield of pulp was calculated on fruit weight basis in this method.

Hot extraction

In hot extraction method the pulp was soaked in water and heated to a temperature of 70°C for 10 minutes. The ratio was retained as flesh: water as 1:2, followed by soaking for 6 hrs. After soaking the mixture was homogenized and filtered to obtain fine pulp.

5.3.2. Pre-treatment for the Tamarind fruit Pulp

Pre-treatments used to prevent darkening and improve quality of the fruit Pulp. The destruction of pathogens that could cause food borne illness, like *Escherichia coli* O157:H7, Salmonella species, and Listeria mono-cytogenes also achieved during Pre-treatments.

5.3.2.1. *Sulfuring*

Sulfuring is an old method used for pre-treating most of the fruits and Pulps. In this method the sublimed sulfur is ignited and burned in an enclosed box with the fruits or pulp. The sulfur fumes penetrate the fruit pulp and retarding spoilage and darkening of the fruit and Pulps.

5.3.2.2. *Sulphite Dip*

The long-term anti-darkening effect was achieved while using Sulphite dips. It will react more quickly and easily. Normally Food grade or Reagent grade (pure) Sulphite dips such as Sodium bisulphite, sodium sulphite or sodium meta-bisulphite can be used.

5.3.2.3. *Ascorbic Acid*

Ascorbic acid combination with water is a harmless way to prevent fruit browning. Its protection does not last as long as sulfuring or sulphating. One teaspoon of powdered ascorbic acid is equal to 3000 mg of ascorbic acid in tablet form. In this method fruit or pulp is kept in the solution for 3 to 5 minutes. Afterwards the fruit or pulp are drain well and place on dryer trays.

5.3.2.4. *Fruit Juice Dip*

The fruit juice with high vitamin C can also be used as a pre treatment, though it is not as effective as pure ascorbic acid. Juices high in vitamin C include orange, lemon, pineapple, grape and cranberry and each juice enhances its own Color and flavour to the fruit under pre treatment.

5.3.2.5. *Steam Blanching*

Steam blanching also helps retain color and slow oxidation. However, the flavour and texture of the fruit or pulp will get change.

5.3.3 Drying methods for the Tamarind Fruit leather

Drying is one of the oldest methods of food preservation techniques and is the most commonly used technique in the food processing industry. Two processes which take place during drying are heat transfer to the product from the heating source and mass transfer of moisture from the interior of the product to the surface. The basic principle of drying is to reduce the moisture content of the product to a level that prevents deterioration within a certain period of time.

For the drying process care must be taken to maintain a steady thickness of product, otherwise moisture patches will cause the fruit leather to rip when it is removed from the drying trays. Common drying methods used to dry the fruit leathers are oven-drying (including convection/fan forced), sun-drying, electric cabinet drying and off-the-shelf food dehydrators (Raab and Oehler, 1999). Both cabinet and oven drying are reported to produce higher quality leathers with cabinet dried leather being more acceptable (Raab and Oehler, 1999).

Open air sun drying is not suited to large scale production. The disadvantages of sun drying are lack of ability to control the drying operation, the slow drying process, dependent on weather condition, high labour costs, insect infections and the exposure to environmental contamination.

Desirable or undesirable chemical or biochemical reactions may lead to changes in color, texture, odour and nutritional properties of the final product during drying process. It was reported that solar and oven dried leathers resulted in a greater loss of color than cabinet dried fruit leather. Improper drying may also lead to physical changes, such as shrinkage, puffing and crystallization of the product. Fruit leathers dried at higher temperatures and for shorter drying times have been found to be darker. The temperatures greater than 60°C caused a rapid increase in non-enzymatic browning. The aroma of solar dried leather was generally disliked and had significantly lower acceptability scores compared to cabinet and oven dried leather ($P < 0.05$). The aroma for both cabinet and oven dried leather was acceptable. The high aroma acceptability scores for cabinet and oven dried leathers could be attributed to the short drying times (6 - 18 hours) used, as opposed to 72 hours for solar drying.

5.3.4. Processing of Tamarind fruit leather

For the processing of tamarind fruit leather preparation fully ripe tamarind fruit Pulp, Sugar (10-15% the pulp weigh), lemon juice or citric acid (2 spoons per kg pulp), and Sodium of potassium meta bisulphite (it is optional, 2 g per kg pulp) are used. The Processing steps are shown in Fig. 5.7.

5.3.5. Candy

Sweetened tamarind fruit is made by peeling whole ripe fruits and pouring boiling (62°C) sugar syrup over the fruits and placing in a deep enamel basin. Boiling syrup is prepared by mixing three parts sugar with one part water. After soaking for three days, the fruits are drained of the old syrup and again covered with freshly prepared syrup. The process is repeated until the fruits are sweet enough. The sweetened fruits are arranged on bamboo racks, dried in the sun and covered with a wire screen to keep off flies. Final drying is done in an oven at a low temperature.

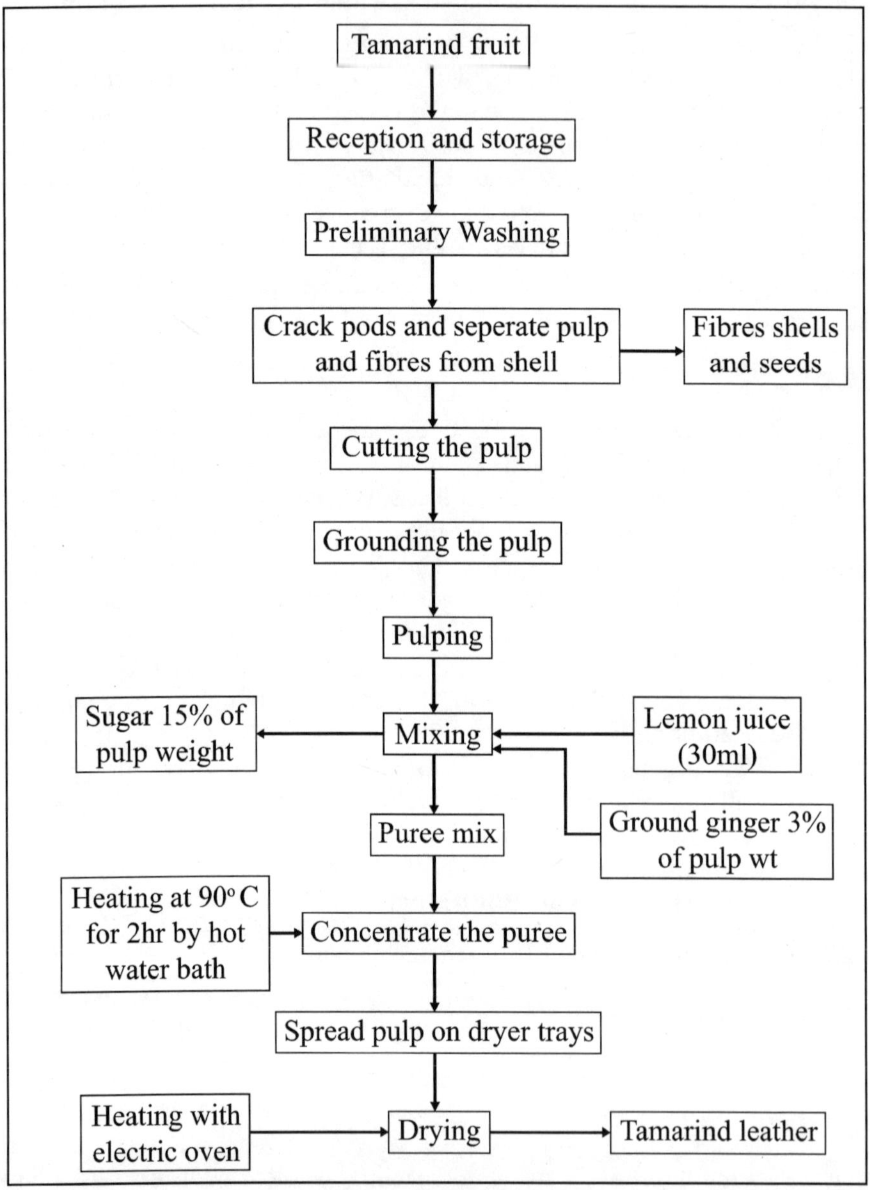

Fig. 5.7. Process flow chart for Developing Tamarind Fruit Leather

5.3.6. Tamarind Ketchup

Wash tamarind pulp in clean water, then boil in clean water and extract the tamarind puree. Mix with 10 per cent sugar, 1 per cent salt and cook on medium heat. Then remove from heat and mix with the spices. Boil the edible oil and put sliced ginger, small pieces of garlic and chili and cool down the product before packing. In the Figure 5.8 Processing flow chart for Tamarind Ketchup is given.

Washing the pulp

↓

Extraction of Tamarind Pulp Puree

↓

Adding 10% sugar, 1% salt

↓

Adding other spices

↓

Mixing sliced ginger, small pieces of garlic and chilli powder

↓

Cooling

↓

Packing

↓

Storage

Fig. 5.8. Processing of Tamarind Ketchup

5.3.7. Tamarind pickle

Pulp is used commercially to prepare tamarind pickle. Pickles are hot, spicy and have a salty-sour taste, and can be preserved for several months. Preservation is due to the presence of salt, increased acidity and spices. The preparations of pickles are simple and can be done at a household level. When making pickles, fresh mature tamarind fruits are selected and soaked in clean water for 12 hours and the pulp is separated. For each kilogram of tamarind, one kilogram of sugar is added and boiled, stirring the mixture continuously. Then mixtures of spices are added. Spice mixture for one kilogram of tamarind should include the following: 40 g coriander, 50 g cumin seed, 30 g black cumin, 3-4 cloves, 3-4 pieces of cardamom, 3-4 pieces of cinnamon, 10-12 chillies, 30 g salt, 250 ml mustard oil and 30 g black pepper. These are ground into a paste and added to the boiling tamarind and mixed thoroughly. The product is then packed into pre-sterilised glass jars or plastic bags and allowed to cool at room temperature. The pickle can be stored in small polythene bags or in clean jars and sealed. It will keep well if stored in a cool place. Care should be taken to prevent air bubbles becoming trapped in the pickle and to ensure that a layer of oil covers the product.

5.3.8. Champoy

Tamarind fruits may also be processed into balls or 'champoy' which is a common tamarind product in the Philippines. Toone cup of pulp with seeds are added two cups of boiled and mashed sweet potato, two cups of sugar, one-eighth cup of salt and one cup of water. The mixture is cooked in a moderate fire with constant stirring until it becomes thick and may be shaped into balls. The balls are rolled in sugar and wrapped in cellophane.

5.3.9. Ade

This is a refreshing tamarind drink prepared in the Philippines and some tropical American countries by mixing ripe pulp with sugar and water until the desired taste is attained. The simple home method of making Ade is to shell the fruits, placing them in 2-3 litres of water and allow it to stand for a short time, then adding a tablespoonful of sugar and shaking vigorously. For a richer beverage the shelled tamarind should be covered with hot sugar syrup and allowed to stand for several days and finally diluted to the required strength with iced water and strained. Sometimes spicy seasonings, such as cloves, cinnamon, ginger, pepper or lime slices, are added to improve the taste (Morton, 1987). In Brazil, Ade is prepared by covering the shelled fruits in cold water for 10-12 days, then the seeds are removed and strained and a cup of sugar is added to every two cups of pulp. The mixture is boiled for 15-20 minutes and placed into glass jars topped with paraffin and diluted as desired. It is a popular drink in the tropics of America and it is bottled in carbonated form in Guatemala, Mexico and Puerto Rico.

5.3.10 Value-added Products from Tamarind

NCL scientists have discovered a process to recover tartaric acid, pectin and fruit sugar from tamarind and have been awarded a US patent (US 5994533) for the same. These products are widely used in the pharmaceutical sector and in the foods industry to produce jams and jellies, dairy products and baked goods. The NCL process provides an economically viable alternative source for tartaric acid, the only other route to which is through the isolation of argols during manufacture of wine. Commercial exploitation of this discovery could insulate the market price of tartaric acid from cylic fluctuations in grape yields and, reduce or eliminate the current import bill for the country (estimated to be around Rs. 500 million for 1200 tonnes of tartaric acid). NCL's patented process involves sequential isolation and recovery of a mixture of glucose and fructose, pectin and potassium bitartarate (cream of tartar) from tamarind pulp. The viability of the process based on the yields at a 500 gm-scale has been established.NCL has filed for four patents based on this know-how of which one has already been granted.

5.3.11. Tamarind juice powder

The tamarind after dehulling is soaked in hot /cold water overnight. After agitation and expulsion of seed, clear tamarind pulp extract is obtained by filtration process. The tamarind pulp extract is dehydrated in to powder either by freeze drying, spray drying, foam mat drying or by drum drying techniques. Lewis and Neelakantan (1964) recommended the drum drying of tamarind pulp and moulding the tamarind juice powder into cubes for industrial uses. The recommended specification of the tamarind juice powder is given in the Table 5.1.

Table 5.1. Composition of Tamarind Juice Powder

Composition of Tamarind Juice Powder	Quantity (%)
Moisture content	8.5
Total ash	3.0
Acid insoluble ash	0.5
Crude fibre	3.5
Added starch	35
Tartaric Acid	25

(IS 13895:1994)

Technological flow chart for the preparation of tamarind juice powder

Tamarind

↓

Dehulling & Defibring

↓

Soaking

↓

Pulping → Fibre, seed, Cellulose residue

↓

Pulp extract

↓

Filtering

↓

Freeze drying/Foam mat drying/Spray drying/Drum drying

↓

Tamarind juice powder

↓

Packing

↓

Storage

5.3.12. Champoy

Tamarind balls or Champoy is used traditionally in Philippines. Tamarind pulp is mixed with salted mashed sweet potato and the semi-solid is balled and coated with sugar. Thailand sweet tamarinds are simply harvested and packed in cellophane or butter paper and exported to fetch a lucrative foreign exchange (Gunasena and Hughes, 2000).

5.3.13. Tamarind Jam

Tamarind pulp is rich in pectin ranging from 2-4 per cent. Tamarind pulp pectin resembles apple pectin. Hence, the semi-solid product from fruits like jam can be prepared from tamarind due its pectin quality. Tamarind pulp is boiled with water to get pulp extract and two part of sugar is added to one part of tamarind extract (Gunasena and Hughes, 2000). After achieving the end point, jam is filled hot into sterilized bottles and cooled. Wax coating to avoid fungal contamination is recommended before lidding the bottles.

Technological flow chart for the preparation of the Tamarind Jam

Tamarind fruit
↓
Boiling with water
↓
Pulp extraction → Seed removal
↓
Addition of Sugar
↓
(1:2 – Pulp: Sugar)
↓
Boiling (till reaching half of original volume)
↓
Judging of end point
↓
Filling hot into sterilized bottles
↓
Cooling
↓
Waxing
↓
Capping
↓
Storage

5.3.14. Tamarind pickle

Pickling is an age old practice of preserving fruits and vegetables. In pickling, various techniques such as preservation with salt, oil, sugar, preservatives or combination of salt and oil are followed. Tamarind pickles are acidic in nature and hence, free from bacterial contamination and preferred by everyone due to its spicy taste. Fungal contamination is a quite common spoilage caused in most of the pickles. Hence, the recommended practices to avoid fungal contamination are

1. The use of plastic spoons to scoop away pickles,
2. Storage of pickles in glass jars,
3. Shaking of pickle at least twice in a day and
4. Keeping pickles under sun for moisture migration

Technological flow chart for the preparation of tamarind pickle

Tamarind

↓

Boiling with water

↓

Pulp extraction

↓

Addition of sugar

↓

Boiling

↓

Heating of oil

↓

Cooling of oil

↓

Mixing of powdered spices in quarter amount of oil (spice paste)

↓

Mixing of spice paste with boiling sugar and pulp mixture

↓

Filling hot into sterilized bottle

↓

Cooling

↓

Capping

↓

Storage

Tamarind pulp boiled with water and pulp is extracted after seed removal. The pulp extract is mixed with equal amount of sugar and boiled continuously. Meanwhile, the spices are powdered and added with little amount of oil to make paste. The spice paste is mixed with the boiling sugar and pulp mixture. The pickle is mixed into the sterile bottle and stored after capping (Gunasena and Hughes, 2000).

5.3 LEAVES AND FLOWERS

Tamarind leaves are used for folder and are relished by cattle and goats, although they are rarely used for this purpose in home gardens as harvesting reduces fruit yield. However, tamarind trees growing in woodlands are often eaten by wild animals, such as elephants in Zambia. The leaves, flowers and immature pods of tamarind are also edible. The leaves and flowers are used to make curries, salads, stews and soups in many countries, especially in times of scarcity. Young leaves of tamarind are used as a seasoning vegetable in some Thai food recipes because of their sourness and specific aroma (Cronel, 1991). In Plate 5.6 Tamarind Flowers and Leaves are given. In Zimbabwe, the leaves are added to soup and the flowers used in salads. The children in Gambia mix the acid leaves with gum from Fig. trees to make a 'chewing gum'.

Fig. 5.9. Tamarind Flowers and Leaves

REFERENCES

Baragano de Mosqueda, M. (1980). Technology of clarified tamarind juice. In Tropical and Subtropical Fruits. (Eds) Nagy, S. and Shaw, P.E. *AVI Publishing,* Westport, Conn., 375.

Bueso CE. (1980). Soursop Tamarind and Chironka, In: Tropical and Subtropical Fruits, by S Nagy and PE Shaw (Eds), *AVI Publishing,* Westport, Conn, 1980, pp. 375.

Chapman, K.R. (1984) Tamarind. In Tropical Tree Fruits for Australia. Compiled by P.E. Page. Queensland Department of Primary Industries. Information series Q 183018, Brisbane: 83-86

Gunasena, H. P. M and A. Hughes (2000). Tamarind. Published by International Centre for Underutilized Crops. University of Southampton. Southampton. UK. Pp:1-188.

Gunasena, H.P.M. (1997) Domestication of agroforestry trees in Sri Lanka. Proceedings of the Regional Workshop on Domestication of Agroforestry Trees in Southeast Asia. *A publication of Win rock International* in collaboration with ICRAF. Yogyakarta, Indonesia: 49-53.

Lewis, Y.S. and S. Neelakantan. (1964). The chemistry, biochemistry and technology of tamarind. *Journal of Science and Industrial Research,* **23:** 204.

Morton, J. (1987) Fruits of Warm Climates. Miami FL: 115-121

Nagalakshmi, S. and Chezhian, N. (2004) Influence of type of packaging and storage conditions on Tamarind (*Tamarindus indica*) pulp quality. *Journal of Food Science and Technology,* **41(5)**: 586-590.

Puranaik, J., Nagalakshmi, S., Balasubrahmanyam, N. and Sakaracharya, N.B. (2004) Packaging and storage studies on Tamarind (*Tamarindus indica.* L) Pulp. *Journal of Food Science and Technology,* **14(3)**: 348-351

Purseglove, J.W. (1987) Tropical Crops. Dicotyledons, *Longman Science and Technology,* 204-206

Raab. C. and Oehler. N. (1999) Making dried fruit leather. Fact Sheet 232, pp 1-4, Oregon State University Extension Service.

Shankaracharya, N.B. (1998) Tamarind - Chemistry, Technology and Uses - a critical Appraisal. *Journal of Food Technology,* 35(3): 193-208.

Vernon-Carter, E.J., Espinosa-Paredes, G., Beristain, C.I. and Romero-Tehuitzil, H. (2001) Effect of foaming agents on the stability, rheological properties, drying kinetics and flavour retention of tamarind foam-mats. *Food Research International,* **34(7)**: 587-598.

■■■

Chapter 6

BY PRODUCTS OF TAMARIND

Miss. M. Sangareswari Nagajothi* & Dr. P. Sudha**

**Ph.D. Scholar, Department of Silviculture,*
Forest College and Research Institute, Mettupalayam
***Assistant Professor (Food and Agricultural Processing Engineering),*
Department of Silviculture, Forest College and Research Institute, Mettupalayam

Tamarind has multiple uses and every part of the tree (leaves, flowers, barks, twigs, fruit and seed) is useful. Some are well known and some need further development and awareness. The unique sweet and sour taste of fruit makes awesome taste foods. The main by product of tamarinds is the seed. The tree produces brown pod-like fruits, which contain pulp and hard-coated seeds. The fully formed each seed is enclosed in a parchment and are hard, glossy-brown, square shaped with rounded corners and edges (Fig. 6.1).

Fig. 6.1. Tamarind seeds

A mature tree may annually produce 150 to 225 kgs of fruits, of which contains 30 to 55 per cent of pulp, 11 to 30 per cent of shells and fibre and 33 to 40 per cent of seeds. The present productivity of the tamarind seed is poorly estimated compared to the actual productivity. The reason behind is a vast quantity of tamarind seed is wasted at farm gate itself due to the ignorance of farmers, lack of availability of good harvesting and processing techniques, and lack of recommended silvicultural practices and less developed marketing channels for tamarind seed as well as pulp.

6.1. COMPOSITION OF SEEDS

Following the estimation of the composition of seeds and evaluation of its properties, tamarind seeds are potential sources of food or food ingredients (Marangoni *et al.*, 1988). Tamarind seed comprises the 20-30 per cent seed coat or testa and the kernel or endosperm is about 70-75 per cent (Fig. 6.2). The chemical composition and nutritive value of tamarind seeds (17.1–20.1% protein 6.0–8.5% fat; 65.1–72.2% carbohydrates; 0.7–4.3% crude fibre and 2.3–3.2% ash) (Bose *et al.*, 1954; York *et al.*, 1993; Siddhuraju *et al.*, 1995; Patil and Nadagouder, 1997). Tamarind seed is the raw material used in the manufacture of tamarind seed kernel powder (TKP), polysaccharide (jellose), adhesive and tannin from testa. The seeds are also used for other purposes and are presently gaining importance as an alternative source of protein, rich in some essential amino acids. Presently, Tamarind kernel is a considered as a kind of waste that is not yet used optimally. As natural hydro-colloid source, for it is containing starch and gum, the kernel can be changed in powder form to increase its utility.

Fig. 6.2 Tamarind seed powder

6.2. TAMARIND KERNEL POWDER

In processing tamarind powder, seeds must be cleaned before commencement of processing. The production process consists of cleaning the tamarind seeds by passing through a chaff separator and destoner. Here, contaminant material such as twigs, chaff, fibre, mud, and stones are removed. Then kernel shell must be separated. There are two methods to remove seed coat or testa.

(a) Decortication

One manner to separate testa is by roasting and another way is soaking in cold water for 24 hours, but roasting is a simple and time saving method for removing seed testa. Once seed coat is removed the kernel is powdered by pulveriser then it can be used for many purposes. The cleaned seeds are then conditioned by addition of water, passed through a roaster to loosen the skin and then ground in a hammer mill to 200 mesh particle size and passed through a vibratory sieve to separate the fibre from starches. Presently machineries are available for tamarind seed decortications.

(b) Tamarind seed decorticator

Tamarind seed decorticator (Fig. 6.3) developed at Forest College and Research Institute, Mettupalayam works on the principle of impact and shear. This machine has the capacity of 200 kg/hr operated by 2 HP motor. By passing the decorticated seeds the broken testa is separated in an oscillating screen separator. The efficiency of this decorticator depends on proper roasting of seeds prior to feeding in the decorticator.

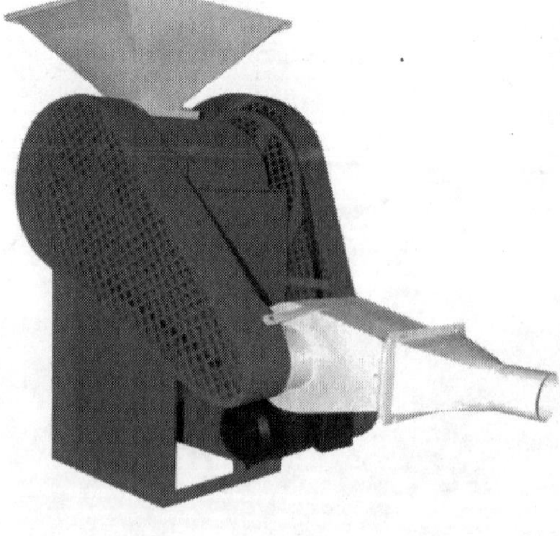

Fig. 6.3. Tamarind Seed Decorticator

6.2.1. Seed powder

Tamarind seed powder is a rich source of Polysaccharide, which contains monomers of galactose, xylose and glucose sugars present in a molar ratio of 3:1:2, which constitutes about 65 per cent of the seed components. The Xylose is very crucial sugar of tamarind seed and used for xylitol production. Bioavailability of some drug has been reported to significantly improve the use of Tamarind seed polysaccharide.

Kernel protein is rich in lysine, glutamic acid, aspartic acid, glycine, leucine and potassium, but deficient in sulphur-containing amino acids (Bhattacharya *et al.*, 1993). Dehusked tamarind seeds have been found to be a rich source of pectin which is the jelly-forming constituent of many fruits, vegetables, seeds, etc. (Kumar, 1997). Proper utilization can give an impetus to the jam and jelly industry which until now was dependent upon the imported jelly powder, and can also lead to the development of various other industries which use pectin as one of its raw materials. The powder, commercially known as tamarind kernel powder (TKP), is found to be extensively used as a sizing material in the textile industry as well as in the food industry (Rao and Subramanian, 1984; Bal and Mukherjee, 1994; Patil and Nadagouder, 1997). The sizing properties of TKP are by presence of a polysaccharide (called jellose) to the extent of 6 per cent. Other reported constituents are proteins, fibre, fat and inorganic salts and some free sugars and tannins. The jellose is also much used in confectionery, especially in the United States, and some European countries. Its use has been recommended in preparing jujubes, as a stabilizer in ice creams and mayonnaise (Patil and Nadagouder, 1997).

Xyloglucan is a major structural polysaccharide in the primary cell wall of higher plants. It shows properties like high viscosity, broad pH tolerance and adhesively. This led to its application as stabilizer, thickener, gelling agent and binder in food and pharmaceutical industries. In addition to these other important properties like non-carcinogenicity, mucoadhesivity, biocompatibility, high drug holding capacity and high thermal stability. Which is present in tamarind kernel powder, this led to its application as excipient in hydrophilic drug delivery system.

6.2.2. Characters of tamarind seed polysaccharide

Tamarind Seed Polysaccharide (TSP) is soluble in ethanol, methanol, acetone, ether and cold water, it is insoluble in organic solvents, but it yields a highly viscous colloidal solution or a viscous gel at temperatures above 85°C and it is completely dissolved in hot water (Fig. 6.4). It possess the properties such as high viscosity, adhesivity, non-carcinogenicity, broad pH tolerance and biocompatibility. It is also a potential emulsifier, nontoxic and non-irritant agent having haemostatic activity. It includes the high swelling index high thermal stability and high drug holding capacity, making it a suitable excipient for drug delivery system. Tamarind seed polysaccharide increases the viscosity showing muco adhesive, muco mimetic and bio adhesive activities. Tamarind kernels also have the property of forming films which have high tensile strength and flexibility, making it a good excipient for ocular preparations (Table 6.1 and 6.2).

Fig. 6.4. Tamarind seed polysaccharide

Table 6.1. Particle size distribution of tamarind seed polysaccharide

Size (μ m)	Number of particles
0-30	10
30-60	52
60-90	258
90-120	170
>120	18

Source : Phani Kumar *et al.* (2011)

Table 6.2. Characterization of tamarind seed polysaccharide powder

Property	Results
True density (g/cc)	1.015 ± 0.04
Tapped density (g/cc)	0.781 ± 0.02
Bulk density (g/cc)	0.651 ± 0.04
Angle of repose (*)	29.50 ± 0.12
Compressibility index (%)	16.64 ± 0.04

Property	Results
Hausner ratio	1.02 ± 0.09
pH	6.81± 0.21
Swelling index (%)	1700 ± 0.56
Surface tension (dynes/cm)	83.26 ± 0.62
Moisture content (%)	8.10 ± 1.23
Water retention (%)	20.00 ± 1.34
Melting point	240-260

Source : Phani Kumar *et al.* (2011)

6.2.3. Storage properties of tamarind seed powder

TSP can be stored for a long time at 40°C and 75 per cent relative humidity. After 6 months of storage also there are no much changes in their properties. The microbial studies of tamarind seed polysaccharide proved that tamarind seed polysaccharide does not support microbial growth and is free from all the pathogen organisms (Table 6.3 and 6.4).

Table 6.3. Tamarind seed polysaccharide after storage of 6 month at 40°C/75% RH

Condition	Duration (Months)	Parameters		
		Description	Moisture content	pH
Initial	Initial	Cream, odourless Hard and rough and free flowing powder	8.10	6.81
40°C/75 % RH	6	Cream, odourless Hard and rough and free flowing powder	8.66	6.95

Source : Phani Kumar *et al.* (2011)

Table 6.4. Microbial load after storage of 6 months at 40°C/75% RH

Condition	Duration (Months)	Parameters					
		Total bacteria	Total fungal count	Pathogens			
				Staph. aureus	*E. coli*	*P. aeroginosa*	*Salmonella*
Initial	Initial	32 CFU	28 CFU	Absent	Absent	Absent	Absent
		39 CFU	33 CFU	Absent	Absent	Absent	Absent

Source : Phani Kumar *et al.* (2011)

6.3. TAMARIND SEED GUM

Tamarind seed gum, a crude extract of tamarind seeds, is rich in polysaccharide, which forms viscous solutions when dissolved in water as many polysaccharide gums extracted from plant materials, which contain glucose, xylose and galactose units, in a molecular ratio of ~3:2:1. Its structure is based on a β (1→4)-D-glucan backbone, substituted at position 6 of the glucopyranosyl units mainly by single α-D-xylopyranosyl residues as well as by disaccharide side chains composed of β -D-galactopyranosyl-(1→2)- α -D-xylopyranosyl residues (Patel *et al.*, 2008). In addition, tamarind seed gum is a high molecular weight polysaccharide (720-880 k Da), which forms viscous solutions when dissolved in water as many polysaccharide gums extracted from plant materials.

6.3.1. Preparation of tamarind gum

Laboratory scale extraction of Tamarind gum is described by Khullar *et al.* (1998). Twenty grams of tamarind kernel powder are added to 200 ml of cold distilled waste and slurry can be prepared. Slurry is poured into 800 ml of boiling distilled water and boiled for 20 min with stirring in a water bath. The resulting thin clear solution should is kept overnight for the protein and fibre to settle down. the Resulting solution is centrifuged at 5000 rpm for 20 min. The supernatant is separated and poured into twice the volume of absolute ethanol with continuous stirring. By pressing the precipitate in between cheese cloth and washing the product with absolute ethanol, diethyl ether and petroleum ether, the material is dried in oven at 50–60°C under vacuum, then ground, sieved and stored for further purification.

6.3.2. Purification of tamarind Gum

Purification can be done by Fehling solution. For purification stir ten grams of tamarind gum with 1000 ml of water in a blender. Separate the suspension and heat with stirring on a boiling water bath for one hour and dilute with Fehling solution (25 ml diluted to 50 ml) with vigorous stirring. The viscous blue gel will be formed and drain with cheese cloth and wash with distilled water (3 x 500 ml). Suspend the gel in distilled water (600 ml) and dissolve by adding 4 ml hydrochloric acid (2 N). Precipitate the polysaccharides with 400 ml ethanol (95%), drain, squeeze on cheese cloth and wash with 100 ml ethanol (50%) until washings gave negative test for presence of chloride. Dry the product under vacuum.

6.3.3. Rheological properties of tamarind gum

Rheological property, especially flow behavior index, of high temperature extracted Gum shows the samples as non-newtonian pseudoplastic type, and with higher consistency index than of room temperature extracted jellose (Fig. 6.5).

Precipitation with ethanol at pH 3, 4, and 5 of both extraction temperature group samples showed non-significant on consistency index (Sukhawanli and Thamakorn, 2014.)

Prepare slurry by adding 20g of Tamarind kernel powder in 200ml of Cold distilled water

↓

Pour slurry into 800ml of boiling distilled water and boil for 20 min with stirring in a water bath

↓

Keep the resulting thin clear solution overnight

↓

Centrifuge the solution at 5000 rpm for 20 min.

↓

Seperate the supernatant and pour it into twice the volume of absolute ethanol with continous stirring

↓

Press the precipitate in between cheese cloth and wash the product with absolute ethanol, diethyl ether and petroleum ether

↓

Dry the material at 50-60° C under vacuum, then ground and sieve and store for further purification

Fig. 6.5. Flow chart for preparation of tamarind gum

6.3.4. Uses of tamarind seed Gum

Presently, it has potential for commercial applications for examples in the pharmaceutical industry for controlling drug release (Sumathi *et al*., 2002) and in the textile printing as a thickening agent [Abo-Shosha *et al.* (2008); Tepparin *et al.* (2011); Ibrahim *et al.* (2010)].

Particularly in the food industry in Japan, refined tamarind seed gum as permitted food additive has been used for modifying texture as a thickening, stabilizing and gelling agents.

- White Tamarind Kernel Powder is used in food products namely jelly, fortified bread and biscuit (Bhattacharya *et al.*, 1991, 1994b).

- It is used in pharmaceutical, cosmetics and insecticidal preparation. Tamarind seed polysaccharide can be considered as an alternative of hyaluronic acid in green cosmetic products (Alessandra *et al.*, 2015).
- It is used as an adhesive in bookbinding, cardboard manufacture and plywood industry, and in sizing and weighing compositions in the leather industry (Daw *et al.*, 1994; Patil and Nadagouder, 1997; Prabhanzan and Ali, 1995).
- Tamarind seed gum reinforced banana fibre composite materials shows good fire retardant and it can be used as false roofing instead of thermo coal (Kiruthika *et al.*, 2012)

6.4. SEED TESTA

The testa is reported to contain 40 per cent water solubles, 80 per cent of which is a mixture of tannin and colouring matter (FRI, 1955). In the production of TKP or the jellose, large quantities of testa are left as a residual by-product (Fig. 6.6).

Testa is used in dyeing and tanning industries. Seed coat (Testa), a by-product of tamarind gum industries used as a safe antioxidant to increase the shelf-life of foods by preventing lipid peroxidation at low-cost (Rao and Srivastava, 1974; Glicksman, 1986; Tsuda *et al.*, 1994, 1995; Sankaracharya, 1998).

Testa can be used as a substrate to grow *Pleurotus florida*, in order to convert organic wastes into biofertilizer (Madhulatha and Pitchai, 1997).

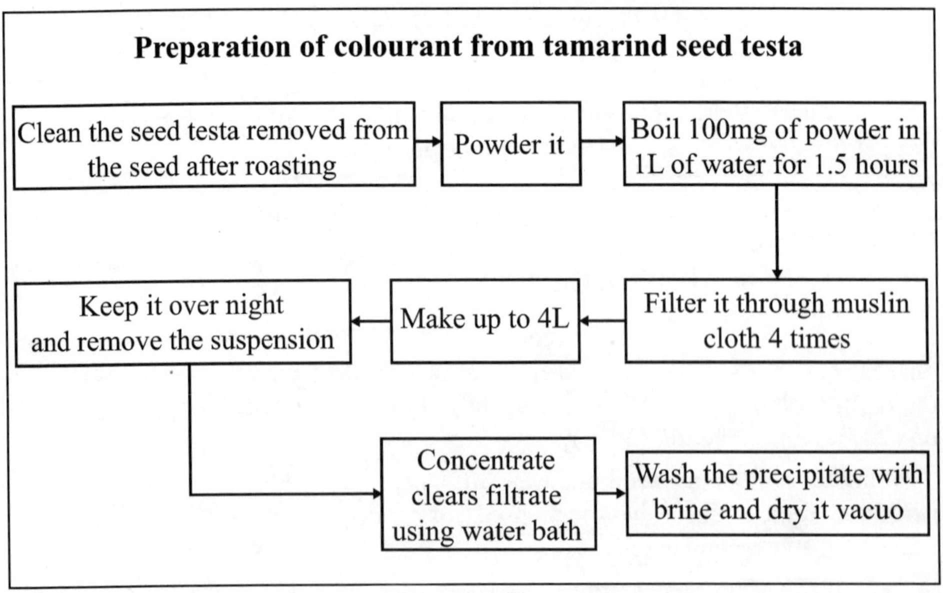

Fig. 6.6. Flow chart for preparation of colourant from seed testa

6.5. SEED OIL

Tamarind seed kernel yields amber-colored oil, free of smell and sweet in taste, which resembles linseed oil. It is used for varnishes, paints and as burning lamp oil (Morton, 1987; El-Siddig *et al.*, 2006), but it is also said to be palatable and of culinary quality (Watt, 1893). Tamarind oil has iodine value below 100 mg/100 g which places it in the non-drying oil group (Ajayi *et al.*, 2006).

6.6. FRUIT SHELL

Tamarind fruit shells have the capacity to remove heavy metals such as Chromium Fluoride and Nickel from aqueous solution. Powdered tamarind shell can be used to treat waste water. It is cheap and easy method for treating waste water. Adsorbent prepared from tamarind fruit shell has a significant capacity for adsorption of Chromium (VI) and Nickel (II) ions from aqueous solution and can be employed effectively as a low cost adsorbent. (Kalnake, 2013). Activated carbon prepared from *Tamarind wood* with zinc chloride activation could be used to effectively adsorb lead (II) from aqueous solutions (Fig. 6.7).

Fig. 6.7. Fruit shell

6.7. LEAVES AND FLOWERS

Tamarind flowers and leaves can be eaten as vegetables and are prepared in a variety of dishes (ICRAF, 2007). The leaves and flowers are used to make curries, salads, and soups in many countries. Small seedlings are eaten as a vegetable. In Thai food recipes young leaves are used as a seasoning vegetable, it gives specific aroma and sourness to food. Leaves are source of vitamin C ß-carotene and minerals particularly potassium, phosphorous, calcium and magnesium. Even tamarind foliage has a high forage value, it is rarely harvested for this purpose because it affects fruit yields. Tamarind has high crude protein content, trees growing in woodlands are often eaten by wild animals, such as elephants or giraffes, for which tamarind is a preferred plant. (El-Siddig *et al.*, 2006).

Tamarind field with apiculture is good revenue source. Flowers are important nectar resource of honeybees in south India. It yields golden yellow colour honey and with slight acidic nature.

Leaves yield red dye which is used in cloths for giving yellow dint in indigo dyed cloths. Leaves and flowers are used as mordant in dyeing. The leaves are used to rear silkworms in India *Anephe* sp. In West Africa *Hypsoides* sp. The silk is very fine and considered superior for embroidery.

6.8. BARK

Tamarind barks and leaves contain tannins. The bark contains up to 70 per cent of tannin and as such has found a place for use in the tanning industry. In tanning hides and in dyeing the bark is used (Morton, 1987; El-Siddig *et al.*, 2006). In Zambia, bark tannins are used in the preparation of ink and for fixing dyes (El-Siddig *et al.*, 2006). The bark is also burnt to make ink in many other African countries. Tamarind twigs are used as 'chewsticks' and the bark is used as a masticatory, alone or as a substitute of lime in betel nut (El-Siddig *et al.*, 2006). The bark yields the alkaloid hordenine (Morton, 1987).

6.9. LAC

Tamarind tree is a host for lac insect (*Kerria lacca*) that deposits a resin on the twigs. This product should be harvested and sold as a stick lac, but it is not considered as important source.

6.10. WOOD

The wood is highly valued in most producing countries although most trees produce little heartwood (Chaturvedi, 1985) and it may be absent altogether in young trees. The heartwood is generally narrow, hard to very hard, fibrous, heavy and strong, and is regarded as very durable. The specific gravity of the wood ranges from 0.80-0.90 g/m^3. It is purplish brown or mottled brown in colour, dark-reddish veined with an irregular outline and radiating ramifications. It resembles ebony (Dalziel, 1937; Eggeling and Dale, 1951; Bolza and Keating, 1972). The sapwood is up to 20 cm wide; light yellow, sometimes red-streaked, rather fibrous and loosely grained, easily perishable (Kurz, 1877; Troup, 1909). The sapwood is liable to insect attack but the heartwood is durable and termite resistant (Timyan, 1996). The wood quality can be correlated to a degree with the cambial activity and development in relation to the climate of growth (Rao and Rajput, 2001). The bark is sometimes used in traditional medicinal preparations. In terms of constituents it lacks phenols and flavonoids (Neetu and Bohra, 2003).

6.10.1. Uses

Tamarind tree is a valuable timber species widely used for manufacturing of tool handles, furniture, charcoal, oil mills, rice ponders and fuel wood. Tamarind wood has many uses, including making furniture, wheels, mallets, rice pounders, mortars, pestles, ploughs, well construction, tent pegs, canoes, side planks for boats, cart shafts and axles, and naves of wheels, toys, oil presses, sugar presses, printing blocks, tools and tool handles and turnery. Heartwood is more durable and it's used in furniture making but due to its hardness seasoning and working is difficult. Tree tends to become hollow in the centre while it gets old. In North America, Tamarind wood has been traded under the name of "Madeira mahogany".

Tamarind wood have good calorific value of 4850 kcal/kg which makes high heat while burning so it is used as fuel wood in brick making. The wood converted as charcoal is valued for making gunpowder. The ash is used to remove hair from animal hides and can be mixed with pulp for cleansing and brightening brass and copper vessels.

6.11. AGROFORESTRY AND LAND USE

Tamarind trees are widely used in India and Sri Lanka as ornamentals and shade trees, planted along avenues, roadsides, in parks and along river banks (Meghwal, 1997) and have become popular in the United States as a tree for indoor ornamentation (NAS, 1979). On a more practical note, tamarind trees have been used as windbreaks due to their resistance to storms, however, there are some disadvantages to using tamarind in this way. This has been demonstrated by Rodriguez *et al.* (1985) when the yield of Valencia orange trees (*Citrus sinensis*) growing adjacent to a tamarind windbreak, decreased by 50 per cent due to competition. Tamarind also provides a heavy shade due to its broad spreading crown. By raising the tamarind trees along field boundaries the crop suffers from less shade and competition. Tamarind is also used as a firebreak in Mysore as no grass grows under the tree due to heavy shade. In some parts of India it is thus planted around forest margins to act as a firebreak (Troup, 1921 and Salim *et al.*, 1998).

Nowadays, Tamarind is used in agroforestry systems in many parts of the tropics due to its multiple uses. In India, many farmers integrate several species, including tamarind, with their agricultural crops and livestock. The increasing integration of tamarind with other trees and crops on farmlands offers a strategy to minimise the risk of crop failure. Tamarind is also an important tree in home gardens in south and southeast Asia. Tamarind side branches are trimmed regularly providing fodder for the animals and simultaneously opening a greater space for the inter-mixed crops in agroforestry systems of Maharashtra (Relwani, 1993).

6.11.1. Intercropping

In Thailand, when sweet tamarind orchards trees are young intercropping with annuals, such as groundnuts, maize, soybean and beans, is practised. In some districts such as Pak Chong in Nakhon Ratchasima province, tamarind is intercropped with custard apple (*Annona* spp.). In the dry season custard apple becomes dormant like tamarind forming a compatible combination (Yaacob and Subhadrabandhu, 1995). Intercropping of tamarind with annual crops can enhance farm income and improve the well being of rural populations, by providing a constant income from the farming system. Once the tamarind tree has passed its juvenile stage and comes into bearing, it will provide a constant supply of fruit, in addition to the produce from the annual crop species. In this way, the farmer obtains more from the same area of land. Tamarind comes into bearing after 10-14 years and for up to eight years of age intercropping is suitable but less so for vegetatively propagated trees which come into bearing at 4-7-10 years. Tamarind leaf extracts, however, have been reported to reduce mitotic activity and induce chromosomal aberrations in *Allium sativum* root meristems (Yadev, 1986). In Southwest Madagascar, the fallen leaves of tamarind, when placed in the planting holes of corn, reduced emergence and grain yield, indicating an allelopathic effect. Mulching with tamarind leaves is therefore, not advisable (Maille, 1991), although traditionally leaves have been used as mulch material for tobacco cultivation in some countries. Because of the allelopathic effect of tamarind, mixed plantations have not been recommended by Gamble (1922) and Yadev (1986), however, in India, to prevent such undesirable effects, tamarind is planted at a wider spacing, commonly 8m x 8m, 8m x 12 m and 12m x 12m (Jambulingam and Fernandes, 1986).

6.11.2. Taungya system

In the forest-village system of Thailand, where taunga is practised in teak (*Tectona grandis*) plantations, interplanted mixtures with tamarind have been successful (Jordan and Gajaseni, 1990; Dagar *et al.*, 1995). The Central Salinity Research Institute in Karnal, India, has verified the advantages of interplanting agricultural crops, including rice, pearl millet and mustard with tamarind, yield reduction was minimal compared to other forest tree species such as Maharukh (*Ailanthus excelsa*) and teak (*Tectona grandis*). In such mixed cropping systems, tamarind also benefits by receiving supplemental irrigation, weeding and other cultural practices (Dagar *et al.,* 1995). There has been a substantial body of research on the allelopathetic effects of tamarind in relation to intercrops and weed control (Channal *et al.*, 2002, 2002a; Parvez *et al.*, 2003, 2003a).

6.12. SOCIO–CULTURAL ASPECTS

In India and the Far East since ancient times, members of the flora and fauna were identified with particular personalities of the Hindu pantheon and worshipped. Tamarind is such a highly sacred and worshipped tree. These beliefs and practices continue even today in certain sections of society, particularly among the rural folk and aboriginal tribes living in the forest. Nature worship and the faith of these peoples have helped to conserve many natural ecosystems in India. These people have preserved many forest trees such as tamarind and sacred groves are conserved in their original form (Singha, 1995). In some localities of India the tamarind tree is considered to be haunted by spirits and is worshipped on a day called 'Amli Agiaras'. Hindus may also tie a tamarind tree to a mango tree before eating the fruits of the latter and in effect 'marry' the species. The natives of India consider that the neighbourhoods in which tamarind trees grow become unwholesome, and that it is unsafe to sleep under the tree owing to the acid they exhale during the night (Grieve, 1995).

According to Morton (1987), few plants survive beneath the tamarind tree due to its heavy shade and there is a superstition that it is harmful to sleep or to tie a horse beneath the tree, due to the corrosive effect that fallen leaves have on fabrics in damp weather. Tamarind is also used to forecast the weather conditions by the tribal societies in Rajasthan, India. When the fruit is longer than usual, splendid weather with abundant rain is forecast promising high yields of grain and pulse crops (Joshi, 1985). Some African tribes also venerate the tamarind tree as sacred. In Malawi, tamarind bark soaked with corn is given to domestic fowl in the belief that, if they stray or are stolen, it will induce them to return home. To certain Burmese, the tree represents the dwelling-place of the rain god and some hold the belief that the tree raises the temperature in its immediate vicinity. In Malaya, a little tamarind and coconut milk is placed in the mouth of an infant at birth, and the bark and fruit are given to elephants to make them wise.

REFERENCES

Abo-Shosha, M.H.; Ibrahim, N.A.; Allam, E.A. and El-Zairy, E.M. (2008). Preparation and characterization of polyacrylic acid/ karaya gum and polyacrylic acid/ tamarind seed gum adducts and utilization in textile printing. *Carbohydrate Polymers,* **74(2)**: 241-249.

Ajayi, I.A., Oderinde, R.A., Kajogbola, D.O. and Uponi, J.I. (2006). Oil content and fatty acid composition of some underutilized legumes from Nigeria. Food Chemistry, 99, 115-120.appraisal. *Jour. Food Technol,* **35(3)**: 193–208.

Bal, S. and Mukherjee, R.K. (1994). Functional and nutritional properties of Tamarind kernel protein. *Food Chem.,* **49(1)**: 1–9.

Bhattacharya, S., Bal, S., Mukherjee, R.K. and Bhattacharya, S. (1993), 'Some physical and engineering properties of tamarind (*Tamarindus indica*) seed. *J. Food Engineering,* **18**: 77–89.

Bhattacharya, S., Bal, S., Mukherjee, R.K. and Bhattacharya, S. (1994), 'Functional and nutritional properties of tamarind (*Tamarindus indica*) kernel protein', *Food Chemistry,* 49, 1–9.

Bhattacharya S, Bal S., Mukherjee R K and Bhattacharya, S (1991). 'Rheological behaviour of tamarind (*Tamarindus indica*) kernel powder suspension', *J. Food Engineering,* 13, 151–8.

Bose, S.M., Swaminathan, M. and Subramanyam, V. (1954). 'Nutritive value of Tamarind seed', *Bull. Cent. Fd. Technol. Res. Inst.,* **3**: 67–8.

Channal, H.T., Kurdikeri, M.B., Hunshal, C.S., Sarangamath, P.A. and Patil, S.A. (2002) Allelopathic influence of tree leaf extracts on greengram and pigeonpea. *Karnataka Journal of Agricultural Sciences,* **15(2):** 375-378.

Chaturvedi, A.N. (1985). Firewood farming on the degraded lands of the Gengetic plain U.P. Forest Bulletin. No. 50, Government of India Press, Lucknow, India.

Dagar, J.C., Singh G. and Singh, N.T. (1995), 'Evolution of crops in agroforestry with teak (*Tectoma grandis*), maharukh (*Ailanthus excelsa*) and tamarind (*Tamarindus indica*) on reclaimed salt-affected soils. *Journal of Tropical Forest Science,* **7(4)**: 623–34.

Dalziel, J.M. (1937) The Usefül Plants of West Tropical Africa. London. Crown Agents for Overseas Governments and Administrations: 612 p.

Daw, Z.Y., EL Gizawy, S.A. and Said, A.M.B. (1994), 'Microbiological evaluation of some local juices and drinks. *Chemic Mikrobiologic Technologic der Lebensmittel,* **16(1–2)**: 8–15.

El-Siddig, K., Ebert, G. and Lüdders, P. (1999). *Tamarind* (*Tamarindus indica* L.): a Review on a Multipurpose Tree with Promising Future in the Sudan. *Journal of Applied Botany – Angewandte Botanik,* **73**: 202-205.

El-Siddig, K., Gunasena, H.P.M., Prasa, B.A., Pushpakumara, D.K.N.G., Ramana, K.V.R., Vijayanand, P. and Williams, J.T. (2006). Tamarind – *Tamarindus indica* L. Fruits for the future 1. Southampton Centre for Underutilized Crops, Southampton, UK, 188p.

Gamble, J.S. (1922) A Manual of Indian Timbers. London Sampson Low, Marston & Co: 866.

Glicksman M (1986), 'Tamarind seed gum'. In: *Food Hydrocolloids*. Ed M. Glicksman, Vol III, CRC Press, Boca Raton.

Grieve, M. (1995) Tamarinds. Modern Herbal Home Page (http://www.botanical.com/botanical/mgmh/mgmh.html). *Electric Newt*: 1-2.

Ibrahim, N.A., Abo-Shosha, M.H.; Allam, E.A. and El-Zairy, E.M. (2010) New thickening agents based on tamarind seed gum and karaya gum polysaccharides, *Carbohydrate Polymers,* **81 (2):** 402-408.

ICRAF – World Agroforestry Centre. ICRAF Agroforestry Tree Database: *Tamarindus indica* L., URL http:// www.worldagroforestrycentre.org (visited on 31.01.2007).

Jambulingam, R. and Fernandes, E.C.M. (1986). Multipurpose trees and Shrubs in Tamil Nadu State (India). *Agroforestry Systems* **4(1):** 17-32

Jordan, C.F. and Gajaseni, J. (1990) Inter planting of *Tamarindus indica* L in teak plantations. Fast growing trees and nitrogen fixing trees. (Eds.) Warner, D. and Muller, P. Proceeding of International Conference. Marburg, Oct. 8-12. 76-81.

Joshi, P. (1985) Weather indicating plants of tribals in Southern Rajasthan. *Botanical Bulletin of India,* **27(1-4):** 100-104.

Kumar, V. (1997), 'Tamarind Seed – A Valuable Source of Commercial Pectin'. Proc. Nat. Sym. on *Tamarindus indica* L, Tirupathi (A.P.), organized by Forest Dept. of A.P., India, 27–28 June, 1997, pp. 192–7.

Kurz, S. (1877) Forest Flora of British Burma. Vol. 1. 414. Office of the Superintendent of Government Printing, Calcutta.

Madhulatha, W. and Pitchai, S. (1997), 'Detanned Tamarind Seed Testa – A New Substrate for Pleurotus', Proc. Nat. Sym. on *Tamarindus indica* L, Tirupathi (A.P.), organized by Forest Dept. of A.P., India, 27–28 June, 1997, pp. 175–7.

Maille, P. (1991) Low-tech leaf mulch experiment in Madagascar: Negative effects of *Tamarindus indica* L. on corn. *TRI News,* **10(2)**: 18-20.

Meghwal, P.R. (1997) Exploiting under-utilised fruits of the arid zone. *Indian Horticulture, October-December:* 26-27.

Morton, J. (1987) *Tamarind* In: Fruits of warm climates, Morton, J.F. (ed.). Miami, USA, p. 115-121.

NAS (1979) Tropical Legumes: Resources for the Future. Washington DC. pp. 117-121

Neetu, T. and Bohra, A. (2003) *In vitro* study of antifungal activity of *Tamarindus indica* against *Aspergillus flavus* and *Fusarium oxysporum*. In: Henry, A., Kumar, D. and Singh, N.B. (Eds.) *Advances in Arid Legumes Research,* Scientific Publishers (India), Jodhpur, India. 442-444

Parvez, S.S., Parvez, M.M., Nishihara, E., Gemma, H. and Fujii, Y. (2003) *Tamarindus indica* L. leaf is a source of allelopathic substance. *Plant Growth Regulation,* 40(2)107-115.

Patel, T.R.; Morris, G.A.; Ebringerova, A.; Vodenicarova, M.; Velabny, V.; Ortega, A.; *et al.* (2008) Global conformation analysis of irradiated xyloglucans, *Carbohydrate Polymers*, Vol. 74, No. 4 , pp. 845-851, ISSN 0144-8617.

Patil, S.J. and Nadagouder, B.S. (1997) 'Industrial Products from *Tamarindus indica' Proc. Nat. Sym. on Tamarindus indica* L, Tirupathi (A.P.), organized by Forest Dept. of A.P., India, 27–28 June, 1997, pp. 151–5.

Phani Kumar, G.K., Battu, G., and Lova Raju, N.S. (2011). Isolation and Evaluation of tamarind seed polysaccharide being used as a polymer in pharmaceutical dosage forms. *Res J Pharma Biol Chem Sci.;* **2(2):** 274-90.

Rao, P.S. and Srivastava, H.C. (1974) 'Tamarind'. In *Industrial Gums*. Academic Press, New York, pp. 370–441.

Rao, K.H. and Subramanian. N. (1984) Nitrogen solubility and functional properties of tamarind seed kernel proteins. In proceedings of the National Symposium on Protein Foods and Feeds, PP. A 67-84. Madras, India.

Rao, K.S. and Rajput, K.S. (2001) Seasonal accumulation of starch in *Tamarindus indica* L. and *Azadirachta indica* A. Juss. plants in forests of Gujarat state. *Phytomorphology.* International Society of Plant Morphologists, Delhi, India: 1, 51-56.

Relwani, L.L. (1993) Tamarind: A multipurpose tree for diverse environments. *The Bharathia Agricultural and Industrial Foundation Journal*, 13(2): January: 10.

Rodriguez, R., Del Valle, N., Arango, W., Torres, R. and Fernandez, M. (1985) Effect of windbreaker plants on the yield of Valencia oranges. *Century Agriculture,* 12(1): 71-80.

Salim, A., Simons, A., Waruhin, A. and Orwa, C. (1998) *Agroforestry Tree Database: A tree species reference and selection guide and tree seed suppliers directory*. International Council for Research in Agroforestry, PO Box 30677, Nairobi, Kenya.

Siddharaju P,, Vijayakumari K, and Janardhanan K, (1995) 'Nutritional and anti-nutritional properties of the under exploited legumes *Cassia laerijata* wild and *Tamarindus indica* L', *Journal of Food Composition and Analysis*. An official publication of the United Nations University, *International Network of Food Data Systems*, **8**(4), 351–62.

Singha, R.K. (1995) Biodiversity conservation through faith and tradition in India. Some case studies. *International Journal of Sustainable Development and World Ecology* (UK), 2(4): 278-284.

Sumathi, S. and Alok, R. (2002) Release behavior of drugs from tamarind seed poly-saccharide tablets, *Journal of Pharmacy and Pharmaceutical Sciences*, Vol. 5 No.1, pp. 12-18, ISSN 1482-1826.

Tepparin, S.; Sae-be, P.; Suesat, J. and Chumrum, S. (2011) Preparation of Tamarind-seed thickener for pigment printing on cotton, *Advanced Materials Research*, Vol. 233-235, pp. 1388-1391, ISSN 1022-6680.

Timyan, J. (1996) Important trees in Haiti. Southeast consortium for International Development. 1634, 1, Street N.W. Suite, 702 Washington DC.

Troup, R.S. (1921) *The Silviculture of Indian Trees* Vol. 11. Leguminosae (Caesalpi-nieae) to Verbenaceae. 7. *Tamarindus indica* L.: Oxford, Clarendon Press: 263-363.

Tsuda T, Mizuno K, Ohshima K, Kawakishi S and Osawa T (1995) Superential carbon dioxide extraction of antioxidative components from tamarind (*Tamarindus indica* L.) seed coat', *Jour. of Agricultural and Food Chemistry*, **43**(11), 2803–6.

Yaacob, O. and Subhadrabhandu, S. (1995) *The Production of Economic Fruits in South East Asia*. Oxford University Press: 133-142.

Yadev, S.K. (1986) Antimitotic and cytological activities of tropical forests. *Tamarindus indica. Journal of Tropical Forestry*, 2(1): 53-58.

York W S, Harvey L K, Guillen R, Albersheim P and Darvill A G (1993), 'Structural analysis of tamarind seed xyloglucan oligosaccharides using beta-galactosidasic digestion and spectroscopic methods', *Carbohyd Res*., **248**, 285–301.

■■■

Chapter 7

TAMARIND FLOWERING

Dr. V. Ponnuswami* & Dr. E. Alli Rani**

**Former Dean (Hort), Emeritus Scientist (ICAR),*
E-mail: swamyvp200259@gmail.com
***Research Associate, Department of Vegetable Crops, HC& RI, Tamil Nadu Agricultural University, Coimbatore – 641 003*

7.1. INTRODUCTION

Of all the fruit trees of the tropics, none is more widely distributed or more appreciated as an ornamental than the tamarind, *Tamarindus indica* L. (syn. *T. occidentalis* Gaertn.; *T. officinalis* Hook.), of the family Leguminosae. Most of its colloquial names are variations on the common English term. In Spanish and Portuguese, it is *tamarindo;* in French, *tamarin, tamarinier, tamarinier des Indes,* or *tamarindier;* in Dutch and German, *tamarinde;* in Italian, *tamarandizio;* in Papiamiento of the Lesser Antilles, *tamarijn.* In the Virgin Islands, it is sometimes called *taman;* in the Philippines, *sampalok* or various other dialectal names; in Malaya, *asam jawa;* in India, it is tamarind or *ambli, imli, chinch,* etc.; in Cambodia, it is *ampil* or *khoua me;* in Laos, *mak kham;* in Thailand, *ma-kharm;* in Vietnam, *me.* The name "tamarind" with a qualifying adjective is often applied to other members of the family Leguminosae having somewhat similar foliage. The tamarind tree has the ability to grow in poor soils because of their nitrogen fixing capability and withstanding long periods of drought which makes them ideal low input, high yielding trees (Klahal *et al.*, 2013).

7.2. ORIGIN AND DISTRIBUTION

Native to tropical Africa, the tree grows wild throughout the Sudan and was so long ago introduced into and adopted in India that it has often been reported as indigenous there also, and it was apparently from this Asiatic country that it reached the Persians and the Arabs who called it *"tamar hindi"* (Indian date, from the date-like appearance of the dried pulp), giving rise to both its common and generic names. Unfortunately, the specific name, *"indica",* also perpetuates the illusion of Indian origin. The fruit was well known to the ancient Egyptians and to the Greeks in the 4th Century B.C.

The tree has long been naturalized in the East Indies and the islands of the Pacific. In all tropical and near-tropical areas, including South Florida, it is grown as a shade and fruit tree, along roadsides and in dooryards and parks. In India there are extensive tamarind orchards producing 275,500 tons (250,000 MT) annually. The pulp is marketed in northern Malaya and to some extent wherever the tree is found even if there are no plantations.

7.2.1. Phenology of Flowering

7.2.1.1. Flowering Season

In Florida, Central America, and the West Indies, the flowers appear in summer, the green fruits are found in December and January and ripening takes place from April through June. In Hawaii the fruits ripen in late summer and fall.

At present, little is known about how climatic conditions alter its flowering and fruiting phenologies. Ignoring variations in climatic conditions and how they affect phenology may lead to erroneous generalizations of timing of phenological events and even misestimating variations in fruit production and inter-ecological range.

In tamarind, terminal vegetative shoots which bear flowers only in the following flowering season are produced annually (Fig. 7.1). Two types of terminal shoots have been observed by Nagarajan *et al.* (1998), short ones with an erect habit and long ones with a drooping habit. This has been identified as a useful character to evaluate genotypes, since terminal shoot length and foliage production are found to be highly correlated. In an inflorescence the flowers only open on alternative days. Flowering and fruiting times vary from country to country and region to region depending on the latitude and altitude. It is reported that production of flowers varies considerably between selections in India. Those with longer vegetative terminal shoots produce more flowers. The range of variation of flowering and floral organs is illustrated in following Table 7.1.

Table 7.1. Variation of flowering and floral organs seen in different cloned selection

Characteristic	NBN1	NBN2	Clone NBN3	RDB Patna	JRK	S.Ed	LSD	CV (%)
Inflorescence length (cm)	4.38	4.85	5.70	4.91	4.91	0.11	0.23	3.50
Inflorescence per branch	7.56	13.69	13.59	16.32	11.92	0.17	0.37	2.18
Flower per inflorescence	13.28	8.05	15.57	17.09	12.21	0.02	0.05	0.30
Flowers per branch	100.47	109.12	212.70	279.80	145.61	2.77	5.87	2.58
Style length (mm)	4.57	4.62	4.74	4.30	4.64	0.06	0.14	2.24
Ovary length (mm)	6.60	6.74	6.61	6.13	6.21	0.04	0.08	0.93

Source: Nagarajan et al. (1998)

Note: NBN, RDB and JRK are tamarind clones identified from Gottipura, Karnataka State, India.

In general, flowering and fruiting of tamarind takes place in the dry season. An extended spell of dry weather may be essential for fruit development, and trees which grow in the humid tropics without this dry spell often do not bear fruit. At higher altitudes shoots grow mainly in spring and flower throughout the summer. In the monsoon climate of East Java the tree changes its leaves towards the dry season (September to November), and some trees may be nearly leafless during this dry period. Shoot growth continues through the rainy season (November to April) during which period flowering occurs.

In Thailand, tamarind sheds its leaves and produces new ones towards the end of the dry season (March to April). Sometimes trees are leafless for a brief period, though they often remain foliated. Spasmodic or incidental shoot growth continues into the dry season but in May the trees are usually quiescent. Trees are often foliated, but may be briefly leafless for short periods in hot, dry localities (Chaturvedi, 1985). Shoot growth continues through the rainy season into the dry season. The new leaves appear in March to April. However, some trees flower as late as July or August in the rainy season. In India early, mid and late flowering types of tamarind have been identified and those with delayed flowering habit are reported to be high yielders (Usha and Singh, 1994). In East Java, the flowers appear from April to June. In Zambia, flowering occurs in the hot dry season (October onwards). North of the equator, flowering is from April to May. Mass flowering is common in tamarind during the flowering season (Fig. 7.1). However, in some trees flowers can be seen at any time during the year and may be due to genotype environment interactions (Jayaweera, 1981; Coronel, 1992; Nagarajan *et al.*, 1997).

Fig. 7.1. Profuse Flowering nature

7.2.1.2. Floral Biology

Flowers are cream yellow in colour, bisexual, herkogamous (Fig. 7.2), five sepals, five petals with odd petal being showy, three stamens fused at the base with filaments incurved towards the ovary base (Fig. 7.2), style simple, stigma is unbranched and pappillate, ovary is superior with 12 to 14 ovules.

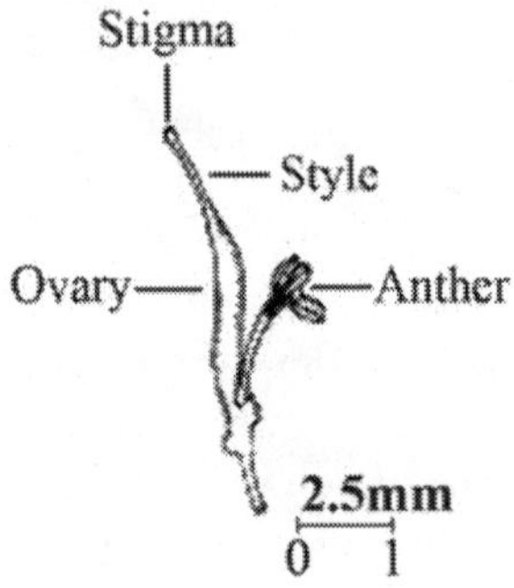

Fig. 7.2. Dissected flower (diagrammatic) of Tamarind

Flowers are borne in lax racemes which are few to several flowered (upto 18), borne at the ends of branches and are shorter than the leaves, the lateral flowers are drooping (Fig. 7.3 and 7.4).

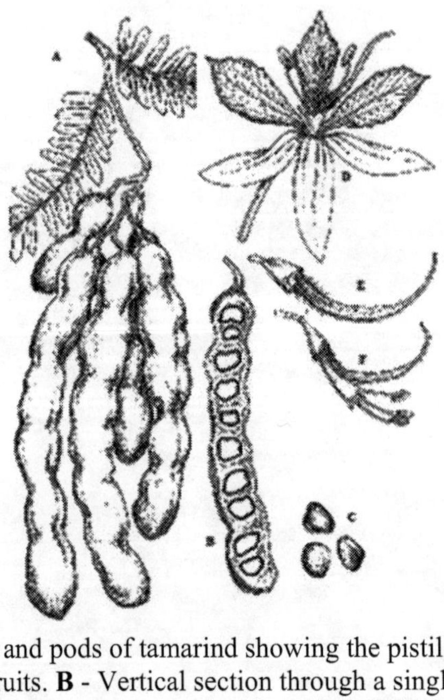

Fig. 7. 3. Flower and pods of tamarind showing the pistil, stamens and seed
A - Cluster of mature fruits. **B** - Vertical section through a single pod/ fruit to show seed arrangement. **C** - 3 seeds. **D** - Flower. **E** -Detail of pistil and stigma. **F** - Detail of corolla tube, pistil and stamens

Fig. 7.4. An inflorescence showing floral parts and sequential development of flowers and fruits

Flowers are irregular 1.5 cm long and 2-2.5 cm in diameter each with a pedicel about (5-) 6 (-10) mm long, nodose and joined at the apex. Bracts are ovate-oblong, and early caduceus, each bract almost as long as the flower bud. There are 2 bracteoles, boat shaped, 8 mm long and reddish.

The calyx is (8-)10(-10) mm long with a narrow tube (turbinate) and 4 sepals, unequal, ovate, imbricate, membranous and coloured cream, pale yellow or pink.

Corolla of 5 petals, the 2 anterior reduced to bristles hidden at the base of the stamina tube. The 3 upper ones are a little longer than the sepals, 1 posterior and 2 lateral, these 3 obovate to oblong, imbricate, coloured, coloured pale yellow, cream, pink or white, streaked with red.

Flowers are bisexual. The colour of the flowers is the same on each tree. They are not mixed. Stamens are 3(-5) fertile and 4 minute sterile ones. Filaments of fertile stamens are connate and alternate with 6 brittle-like staminodes. Stamens are connate and alternate with 6 brittle – like staminodes. Stamens are united below into a sheath open on the upper side and inserted on the anterior part of the mouth of the calyx tube. Anthers are transverse, reddish brown and dehisce longitudinally.

The ovary is superior with few too many (up to 18) ovules. The ovary is borne on a sheath adnate to the posterior part of the calyx tube. It is stipitate, curving upwards and is green with a long hooked style with a terminal subcapitate stigma. Flowers are protogynous, entomophilous and largely cross-pollinated. Flowers are nectiferous, nectar being produced by hairs at the ovary base. Some self pollination also occurs.

The inflorescence racemes are small and drooping with each flower having white or cream petals flecked with pink and about 2 cm in diameter. The bi-sexual flowers exhibit protogynous dichogamy, with flowering normally occurring in spring or summer. Some fruit will be set with self-pollination, however cross pollination is much more effective. Pollination is by honeybees attracted to the nectiferous flowers.

7.2.2. Pollination

Pollination is mostly by honey bees, which is most common among Leguminosae members (Arroyo, 1978). Insect visitation peaks between 08.00 hrs and 11.00 hrs. Anthesis starts at 20.00 hrs and flowers are completely open by 02.00 hrs, anther dehiscence is only by 08.30 hrs in the morning. Stigma is receptive nearly for 48 hours with peak receptivity on the day of anthesis.

Natural or unselected tamarind is a long-lived tree and once it starts bearing, its fruiting capacity increases with age. The productive life of a tree can last for 50-70 years after which it declines. The normal lifespan of the tree is approximately 150 years. It is important to understand the variation in time of the juvenile phase and genotypic and phenotypic variations in flowering, fruit setting and productivity so that domestication can be enhanced.

7.2.2.1. Pollen morphology and pollination vectors

Pollen grains are dimorphic, radially symmetrical, tricolporate, oblate spheroidal in shape and sticky. Nagarajan *et al.* (1997) showed pollen dimorphism in tamarind with two distinct sizes of pollen grains: 40-42 μm and 22-25 μm. This size variation was attributed to resource limitation towards the latter part of the flowering season. Smaller pollen grain counts increased during the later part of the flowering. Pollen sterility is very low in tamarind (less than 2%). Under ambient conditions (37-400°C) pollen viability was observed to be 88% for three days, while pollen stored at 40°C remained 97 per cent viable up to 100 days. Studies carried out in Sri Lanka revealed that tamarind pollen grains are sticky and the flowers produce nectar. Sticky pollens are not efficient for wind pollination. Honey bees (*Apis* spp.) are common visitors to tamarind flowers particularly between 08.00-11.00 hrs and 16.00-18.00 hrs. Observations of honey bee activity on flowers and their visitation patterns suggest that they are effective pollinators. Thus, pollination is mostly by honey bees. Tamarind is insect pollinated due to its floral characters and the presence of copious levels of nectar. Similar observations were also made in India by Nagarajan *et al.* (1998) and Thimmaraju *et al.* (1977). Arroyo (1978) also suggested that honey bees are the most common pollination vector among leguminous species including tamarind. Insect visitation in tamarind peaks between 08.00 hrs and 11.00 hrs which are compatible with stigmatic receptivity and anther dehiscence. Studies in Coimbatore in India observed the foraging of flower visitors. Tamarind flowers mid-May to mid-July but natural fruit set is low: 3-5 per cent compared to 70-90 per cent in controlled experiments. A range of insects belonging to Hymenoptera, Lepidoptera, Hemiptera, Diptera and Thysanoptera visited. Bees are the major pollinators. Domesticated honeybees ensure pollen transfer and fruit set (Sasidharon *et al.*, 2003).

A study was conducted to evaluate the effectiveness of foliar sprays of cycocel (1500 ppm), Ethrel (500 ppm), Triacontanol (20 ml tree^{-1}), IBA (150 ppm), Planofix (100 ppm), micronutrient mixture (0.5%), $ZnSO_4$ (0.5%) + boric acid (0.3%) + $FeSO_4$ (0.5%) and urea (1.5%) on flowering pod set and fruit retention at Tamil Nadu Agricultural University, Coimbatore, exhibited significant effects on flowering, pod set and retention (Ilango and Vijayalakshmi, 2002).

7.2.3. Fruiting phenology

The period from flowering to pod ripening is 8-10 months. Ripe fruits, however, may remain on the tree until the next flowering period. In most of the tamarind producing countries the fruits are harvested from February to March / April, but sometimes the harvesting period may extend to June (Coronel, 1992). In India, fruits are harvested from April to May although in Kerala and other parts of South India fruit collection may be over by the end of February. In Florida, Central America and the West Indies the ripe fruits are available from April to June (Bueso, 1980). In Hawaii, ripening takes place in summer and late fall. In Nicaragua, the harvesting period is early, usually from December to March. In the Philippines, the ripening period is from May to December with a peak in August-October. In Sri Lanka, fruits ripen from January to February to June (Macmillan, 1943; Worthington, 1959). Pods ripen in the spring at high altitudes. In Thailand the fruiting season is December to February. In the Philippines, the fruiting season is from May to December with a peak in August to October (Coronel, 1992). North of the equator, fruit ripening is late in autumn or winter (December to January). In Zambia, fruiting is in the following cool dry season after the main dry season (July to November). In Ethiopia, fruiting is during the dry season, September to April (FAO, 1988).

7.2.3.1. *Studies on Flowering of tamarind under various forest areas in Tamil Nadu*

Through a state forest Department funded Scheme tamarind survey was undertaken at forest areas of different districts of Tamil Nadu and physiological and biochemical analysis were made from soil and leaf samples of two flowering and two non flowering tamarind trees of each location, to investigate the tamarind flowering phenomena in the forest areas and presented in the Table 7.2.

Table 7.2. Study Locations at various districts of Tamil Nadu

S.No	Districts	Name of the Forest Area (Peet)	Range
1.	Theni	Erasakanayakannur	Chinnamannur
		Coodalore	Cumbum
		Kappuvamadai	Gudalore
		Surulipatti	Cumbum
2.	Dindigul	Irandalaparai	Sirumalai
		Kanavaipatti	Natham
3.	Kanyakumari	Aralvaimozhi	Aralvaimozhi
4.	Tirunelveli	Chinnakkadu	Kadayanallur
5.	Tuticorin	Kochanvillai	Tiruchendure
6.	Salem	Kapooth	Karpagam (Salem)
7.	Dharmapuri	Morapoore	Harur
8.	Krishnagiri	Nadhipenda	Burgure
		Kundiyandhal	Burgure
9.	Coimbatore	Mullankadu	Coimbatore
10.	Erode	Bhavanisagar	Bhavanisagar

In these areas, soil and leaf samples of flowering and non–flowering trees were collected and subjected to analysis for physical properties and nutrient composition and the results are furnished in the following the Tables 7.3 and 7.4.

Table 7.3. Physical properties of soil samples (EC, P^H, Texture, Lime status) of tamarind

Location	Ec (dsm^{-1})				p^H				Texture				Lime status			
	Flowering		Non – flowering		Flowering		Non – flowering		Flowering		Non – flowering		Flowering		Non – flowering	
	Tree 1	Tree 2	Tree 1	Tree 2	Tree 1	Tree 2	Tree 1	Tree 2	Tree 1	Tree 2	Tree 1	Tree 2	Tree 1	Tree 2	Tree 1	Tree 2
Natham	0.1	0.2	0.1	0.1	6.3	6.5	6.3	6.2	SL	SL	CL	SL	No	No	No	No
Sirumalai	0.1	0.1	0.1	0.1	6.7	6.5	6.4	6.3	CL	CL	CL	CL	No	No	No	No
Salem	0.1	0.1	0.2	0.1	6.5	6.3	7.2	6.6	CL	CL	CL	CL	No	No	No	No
Harur	0.1	0.1	0.1	0.1	7.6	7.0	6.6	6.3	CL	CL	CL	CL	No	No	No	No
Bargur	0.1	0.1	0.1	0.1	6.8	6.4	6.7	6.8	CL	CL	CL	CL	No	No	No	No
Tiruchendure	0.1	0.1	0.2	0.1	4.6	4.6	4.8	5.1	SL	SL	SL	SL	No	No	No	No
Aralvaimozhi	0.1	0.1	0.1	0.1	5.9	6.0	5.4	5.9	SL	SL	SL	SL	No	No	No	No
Kadayanallur	0.1	0.1	0.1	0.1	6.5	6.3	6.2	6.4	SL	SL	SL	SL	No	No	No	No

Location	Ec (dsm^{-1})				p^H				Texture				Lime status			
	Flowering		Non – flowering		Flowering		Non – flowering		Flowering		Non – flowering		Flowering		Non – flowering	
	Tree 1	Tree 2	Tree 1	Tree 2	Tree 1	Tree 2	Tree 1	Tree 2	Tree 1	Tree 2	Tree 1	Tree 2	Tree 1	Tree 2	Tree 1	Tree 2
Surulipatti	0.1	0.1	0.1	0.1	6.4	6.3	6.5	6.6	CL	CL	SL	CL	No	No	No	No
Coodalore	0.1	0.1	0.1	0.1	6.3	6.3	6.6	6.7	CL	CL	CL	CL	No	No	No	No
Nadhipenda	0.1	0.1	0.1	0.1	6.5	6.4	6.3	6.2	CL	CL	SL	SL	No	No	No	No
Cumbum	0.1	0.1	0.1	0.1	6.9	6.8	6.6	6.6	SL	SL	CL	SL	No	No	No	No
Bhavanisagar	0.2	0.2	0.1	0.1	7.3	7.3	6.7	6.7	CL	CL	SL	SL	No	No	No	No
Coimbatore	0.1	0.1	0.2	0.1	6.9	7.0	7.4	6.8	SL	SL	CL	CL	No	No	No	No
Chinnamanur	0.1	0.1	0.1	0.1	6.7	6.7	6.8	6.8	SL	SL	SL	SL	No	No	No	No

SL – Sandy loam, CL – Clay loam

Table 7.4. Macro nutrient content in (Total nitrogen, phosphorus and potassium) soil samples of tamarind

Location	N (Kg/ac)				P (Kg/ac)				K (Kg/ac)			
	Flowering		Non – flowering		Flowering		Non – flowering		Flowering		Non – flowering	
	Tree 1	Tree 2	Tree 1	Tree 2	Tree 1	Tree 2	Tree 1	Tree 2	Tree 1	Tree 2	Tree 1	Tree 2
Natham	90	105	90	97	T	6	3	T	19	16	22	22
Sirumalai	95	90	99	98	6	5	6	10	13	13	16	16
Salem	101	97	87	106	6	11	5	4	13	13	13	13
Harur	95	97	104	98	T	2	3	4	9	9	9	9
Bargur	87	90	87	90	4	6	4	T	6	6	9	6
Tiruchendure	98	109	104	106	11	3	9	13	16	13	22	16
Aralvaimozhi	106	98	101	97	13	5	5	11	25	25	16	22
Kadayanallur	91	97	101	97	13	6	3	3	31	34	22	25
Surulipatti	106	98	90	94	8	3	4	9	22	19	31	25
Coodalore	90	87	95	87	10	3	5	6	22	22	19	19
Nadhipenda	84	90	98	95	4	4	6	4	6	6	9	9
Cumbum	90	87	94	83	50	14	10	18	6	6	6	6
Bhavanisagar	87	97	95	104	4	18	5	10	6	6	6	6
Coimbatore	91	90	87	84	4	10	18	13	34	34	6	34
Chinnamanur	84	83	91	81	13	5	11	3	34	31	25	25

T - Trace

'Tamarind leaf samples collected from above mentioned districts of different forest areas and the macro nutrient such as total nitrogen, total phosphorus and total potassium were analyzed and the results were presented in the Figs. 7.5, 7.6, 7.7 and carbohydrate : Nitrogen (C:N) ratio, IAA oxidize activity (Table 7.5 and 7.6.) were also observed in tamarind leaf as below. The highest IAA oxidizes activity (µg/g) was recorded in flowering tree. Whereas, lowest in non flowering tree.

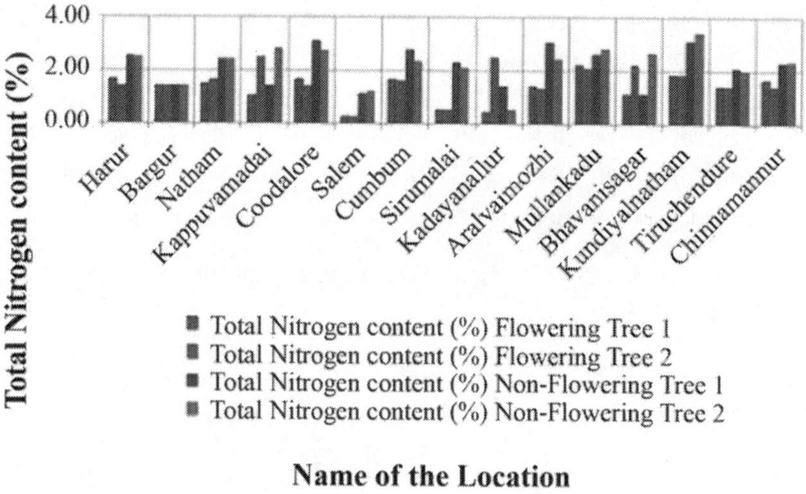

Fig. 7.5. Total Nitrogen content in tamarind leaves (%)

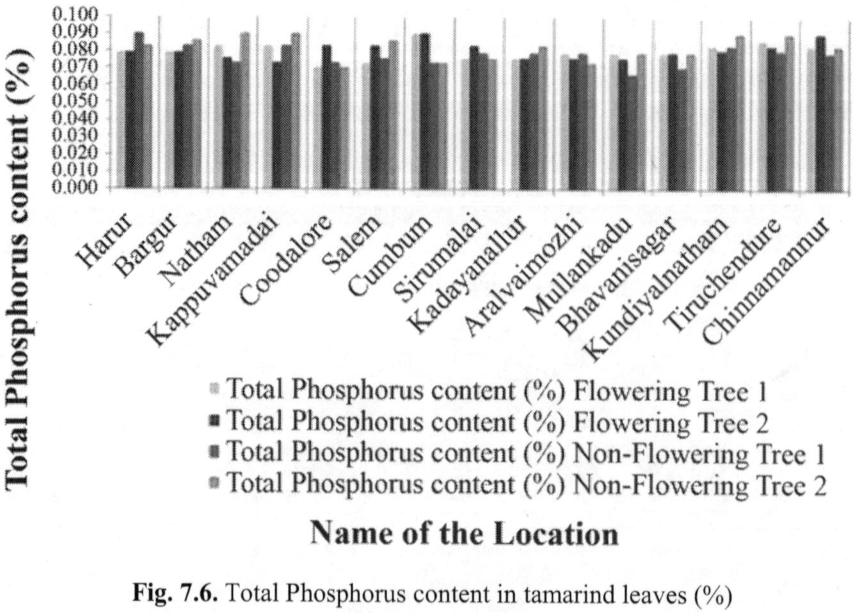

Fig. 7.6. Total Phosphorus content in tamarind leaves (%)

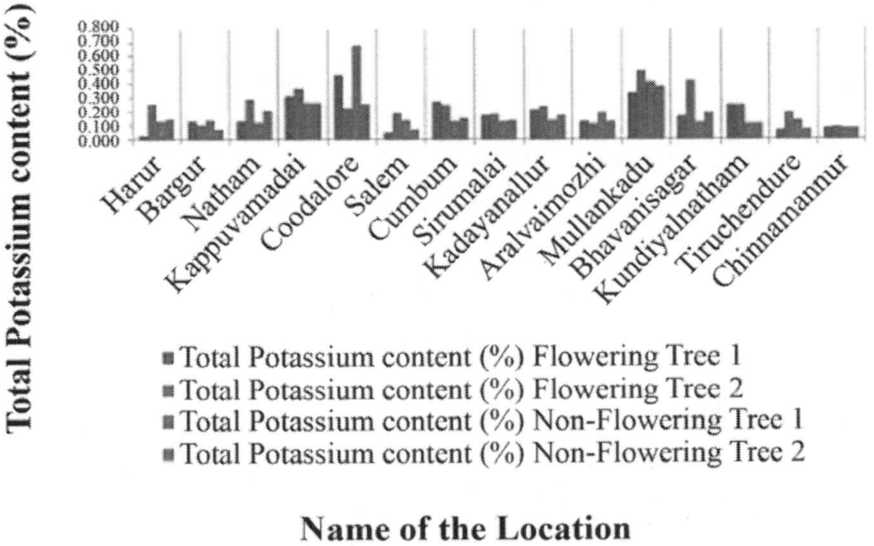

Fig. 7.7. Total Potassium content in tamarind leaves (%)

Table 7.5. Carbohydrate : Nitrogen (C: N) ratio in tamarind leaf samples

Location	Carbohydrate : Nitrogen (C : N) ratio			
	Flowering		Non – flowering	
	Tree 1	Tree 2	Tree 1	Tree 2
Natham	0.007	0.009	0.027	0.028
Sirumalai	0.037	0.040	0.054	0.054
Salem	0.007	0.007	0.032	0.031
Harur	0.011	0.030	0.007	0.004
Bargur	0.006	0.010	0.017	0.019
Tiruchendure	0.046	0.068	0.077	0.071
Aralvaimozhi	0.054	0.008	0.004	0.004
Kadayanallur	0.020	0.020	0.004	0.005
Surulipatti	0.167	0.036	0.064	0.161
Coodalore	0.008	0.055	0.004	0.005
Nadhipenda	0.013	0.023	0.012	0.011
Cumbum	0.014	0.030	0.014	0.006
Bhavanisagar	0.007	0.006	0.028	0.026
Coimbatore	0.008	0.007	0.038	0.040
Chinnamanur	0.006	0.009	0.039	0.038

Table 7.6. IAA oxidize activity

S.No	IAA oxidize activity (μg/g)	
	Flowering	Non – flowering
	123.45	73.90
	97.56	63.23
	135.60	95.54
	79.50	102.36
	167.42	120.30
	157.29	78.30
	89.35	83.76
	107.70	79.29
	114.30	86.65
	94.23	91.40

These studies revealed the following of the main factors that besides the flowering of tamarind trees under the dense forest eco-system.

- Close planting was followed with a spacing of 3 x 3 m.
- The tree height was about 60 to 80 feet
- Variations in trunk, branching and fruits were noticed
- The trunk were less thick, branches were upright and less number of fruits at the top of the canopy in the interior area.
- The canopy is not in an umbrella shape which normally found in tamarind trees.
- The trees in the periphery were flowered and fruit setting was noticed.
- Flowering and fruit setting were also noticed in some of the interior plant, but the numbers of fruits may vary from 5-10 per trees.
- Root spread was more and even it reaches the nearby tamarind trees.
- More weeds were noticed below the trunk of the trees.
- In the peripheral trees horizontal growth was observed towards the outside and vertical growth of the branch was observed in the interior side in one and the same tree. The horizontally grown branches have more number of fruits than that in the vertically grown branches.
- The periphery trees have less flowers and fruit setting was less
- Some of the trees periphery of the canopy only flowers were observed
- Interior trees have less flowering and fruits set and that to found on the top of the canopy.

7.3. CLIMATE AND SOIL

Major tamarind production centers in India are Bihar, Orissa, parts of Maharashtra and Tamil Nadu. Semi-arid tropical climate is best for the commercial cultivation of tamarind. Tamarind can be grown in any area where the temperature reaches 46°C maximum and 0°C minimum. Average rainfall requirement is 500 to 1,500 mm. The optimum altitude required for tamarind cultivation is 1,000 m above MSL (Mean sea-level).

PKM 1 and Urigam tamarind variety released by Tamil Nadu Agricultural University, Horticultural College and Research Institute, Periyakulam and sweet tamarind from Thailand 'Makham waan' and 'Manila sweet' are few important cultivars that need to be exploited for the high tamarind production.

The tamarind is well-adapted humid tropical area with seasonally high rainfall. Young trees are very susceptible to frost, but mature trees will withstand brief periods of 48°C without serious injury. Dry weather is important during the period of fruit development. The tree tolerates a great diversity of soil types, from deep alluvial soil to rocky land and porous, oolitic limestone. It withstands salt spray and can be planted fairly close to the seashore.

7.3.1. Phenology of flowering and fruiting in tamarind

In Coimbatore, the studies on phenology of flowering and fruiting revealed that flowering started from 3rd week of May and extended upto early August with peak flowering during July. The flowering duration was 81 days. Days from bud primardia to anthesis took about 15 days. The pods attained maturity within 235 days. The pod set and pod retention percent were 12.6 and 30.42 respectively (Table 7.7).

7.3.1.1. *Correlation studies*

The weather factors were correlated with flowering and non – flowering nature of various tamarind trees in six districts of Tamil Nadu (Table 7.8) and the results indicated in flowering tree at Theni, Dindigul, Tirunelveli and Dharmapuri districts, the weather parameter maximum temperature had highly significant and positive correlation with Nitrogen, Phosphorus and Potassium content of tamarind leaf and the minimum temperature had a significant and positively correlated with nitrogen and potassium content of tamarind leaf. The rainfall also had a highly significant and positively correlated with nitrogen and phosphorus content.

Whereas, non – flowering tree at Theni district, rainfall had a significant and positive correlation with nitrogen and phosphorus content of tamarind leaf. But in case of Dindigul, Tirunelveli and Dharmapuri districts, temperature and rainfall had a significant and positive correlation with nitrogen and phosphorus content of tamarind leaf observed.

Table 7.7. Flowering season at Coimbatore and Idyarpalayam location in Tamil Nadu

S.N.	Observations	Location	
		Coimbatore	Idyarpalayam
1.	Date of first flowering	22 April – 3 May	15 May – 23 May
2.	Flowering phase		
	Initial	22 April - 17 May	15 May – 14 June
	Peak	18 May – 24 June	15 June – 17 July
	Final	25 June – 28 July	18 July – 03 August
3.	Flowering duration (Days)	99	81
4.	Days from bud primordial to anthesis	15	15
5.	Date of start of pod maturity	June 3rd week (18-24)	June 4th week (25-30)
6.	Pod maturation period (Days)	January 3rd week (222)	February 2nd week
7.	Number of flowers/inflorescence	33	18
8.	Number of inflorescence/tertiary branch	50	36
9.	Length of inflorescence (cm)	13.32	8.5
10.	Pod set (%)	17.4	12.6
11.	Pod retention (%)	37.50	30.42

In flowering tree at Erode and Coimbatore districts, the weather parameter maximum temperature had highly significant and positive correlation with Nitrogen, Phosphorus and Potassium content of tamarind leaf and the minimum temperature had a significant and positively correlated with nitrogen and potassium content of tamarind leaf in erode district. The minimum temperature had a significant and positively correlated with nitrogen, phosphorus and potassium content of tamarind leaf in erode district. The rainfall also had a highly significant and positively correlated with nitrogen and phosphorus content. Whereas, non flowering tree at Erode district, maximum temperature had a significant and positive correlation with nitrogen and phosphorus content of tamarind leaf. In case of Coimbatore district, maximum and minimum temperature had a highly significant and positive correlation with nitrogen and potassium content of tamarind leaf.

The N content of the leaf was highest in the non – flowering trees than the flowering trees, irrespective of the places. The low N content in the flowering trees may be due to utilization of the nitrogen for the production of enzymes which are responsible for the production of energy which helps the tamarind trees to flower.

The potassium content of the leaf was highest in the flowering trees rather than the non - flowering trees, irrespective of the places. The potassium helps in the translocation of photosynthate from the leaves to the flowering branches.

Table 7.8. Correlation co–efficient between weather variables and major nutrient content of leaf in flowering and non – flowering trees at different districts

Weather parameter	**Theni**						**Dindigul**						**Tirunelveli**					
	Flowering tree			**Non - Flowering tree**			**Flowering tree**			**Non - Flowering tree**			**Flowering tree**			**Non - Flowering tree**		
	N	**P**	**K**	**N**	**P**	**K**	**N**	**P**	**K**	**N**	**P**	**K**	**N**	**P**	**K**	**N**	**P**	**K**
Maximum temperature	0.45	0.38	0.49	0.21	-0.19	0.24	0.38	0.29	0.48	0.29	0.22	-0.19	0.33	0.24	0.27	0.27	0.14	-0.20
Minimum temperature	0.37	-0.19	0.34	-0.19	0.23	-0.19	0.27	-0.16	0.35	0.20	-0.18	-0.21	0.21	-0.19	0.33	0.14	-0.21	-0.15
Relative humidity	-0.19	-0.14	0.21	-0.14	-0.21	-0.15	-0.19	-0.21	0.27	-0.25	-0.15	-0.01	-0.17	-.014	-0.17	-0.18	-0.16	-0.17
Rain fall	0.41	0.29	0.19	0.38	0.31	0.24	0.41	0.31	-0.18	0.26	-0.14	-0.23	0.31	-0.18	-0.19	0.31	-0.18	0.24

Weather parameters	**Dharmapuri**						**Erode**						**Coimbatore**					
	Flowering tree			**Non - Flowering tree**			**Flowering tree**			**Non - Flowering tree**			**Flowering tree**			**Non - Flowering tree**		
	N	**P**	**K**	**N**	**P**	**K**	**N**	**P**	**K**	**N**	**P**	**K**	**N**	**P**	**K**	**N**	**P**	**K**
Maximum temperature	0.44	0.35	0.37	0.27	-0.19	0.32	0.35	0.28	0.29	0.29	0.27	0.18	0.52	0.35	0.47	0.25	0.21	0.45
Minimum temperature	0.31	-0.16	0.32	0.18	0.22	-0.16	0.29	0.19	0.27	0.21	0.21	-0.21	0.26	0.30	0.25	0.27	-0.22	-0.22
Relative humidity	-0.17	-0.14	-0.17	-0.17	-0.18	-0.17	-0.16	-0.14	-0.08	-0.16	-0.17	-0.15	-0.21	-0.18	0.19	-0.21	-0.12	-0.21
Rainfall	0.36	-0.13	0.29	0.27	-0.14	-0.07	0.38	-0.07	-0.67	-0.08	-0.14	-0.09	0.26	-0.17	-0.08	-0.15	-0.09	-0.14

N- Nitrogen; P – Phosphorus; K – Potassium

7.4. SUMMARY

Tamarind in general is not so remunerative under dense forest areas in Tamil Nadu or else ware, because of non-flowering nature of the trees. These studies made at Tamil Nadu Agricultural University, under the Tamil Nadu state forest research wing, reviewed that factors like genotypes, climatic factors *viz,* sunshine, temperature, rainfall, wind velocity, humidity, Soil physical, chemical properties, leaf nutrient content, Carbohydrate/Nitrogen (C/N) ratio, endogenous enzyme activity for the critical factors that involved in the induction of flowering of tamarind. These studies in particularly indicated should be optimum C/N ratio, abundant sunlight should focus on the centre of the tree and enzyme activity triggers the auxin and cytokinin activity. The average auxin level and enhance the level of cytokinin in the plants at the time of flowering resulted in flowering. There is a alternate bearing tendency in tamarind like mango. There is proverb in vogue in Tamil Nadu, if one year tamarind flower profusely mango may not flower that much and If mango flower profusely tamarind may not flower that much. But this study in TNAU, the possibility to induce profuse flowering and fruit set in the off year also. If attention is given on the above mentioned riders. If grower adopt a better package incorporating suggestion of the study the tamarind production in waste land and forest areas enhanced with income generation.

REFERENCES

Arroyo, M.K. (1978). Breeding system and pollination biology in Leguminosae. In: Polhill, R.M. and Raven, P.H. (Eds). *Proc. International Legume Conference,* Kew, Royal Botanical Gardens, UK. pp. 723-770.

Bueso, C.E. (1980) Soursop, Tamarind and Chironka. In Tropical and Subtropical Fruits. Nagy, S. and Shaw, P.E. (Eds.), AVI Publishing, Westport, Conn., 375.

Chaturvedi, A.N. (1985) Firewood farming on the degraded lands of the Gangetic plain. U. P. *Forest Bulletin.* Lucknow, India, Government of India Press 50(1): 286.

Coronel, R.E. (1992). *Tamarindus indica* L. In Plant Resources of South East Asia, Wageningen, Pudoc. No. 2. Edible fruits and nuts. (Eds.) Verheij, E.W.M. and Coronel, R.E. PROSEA Foundation, Bogor, Indonesia: 298-301.

FAO (1988) *Fruit Bearing Trees.* Technical notes. FAO-SIDA Forestry Paper 34: 165-167.

Ilango, K. and C. Vijayalakshmi (2002). Effect of some growth regulators and chemic-alson yield and quality attributes in Tamarind (*Tamarindus indica* L). *Orissa Journal of Horticulture,* 30(1): 35-39.

Jayaweera, D.M.A. (1981). Medicinal Plants (Indigenous and Exotic) used in Ceylon. Part 111. Flacourtiaceae-Lytharaceae. *A publication of the National Science Council of Sri Lanka:* 244-246.

Klahal, Kittiya, Janya Pikul, Wancheng Sittikijoythin, Rattanphol and Mongkholrattansit (2013). Thickening agent based on Tamarind seed gum for disperse printing of polyester : *International Textile Conference,* Bangkok, Thailand.

Macmillan, H.F. (1943) Handbook of Tropical Plants. 362. *Medicinal Plants.* Anmol Publications, New Delhi, India.

Nagarajan. B., A. Nicodemus, A. K. Mandal, R. K. Verma, K. Gireesan and N.P. Mahadevan (1998). Phenology and Controlled Pollination Studies in Tamarind, *Silvae Genetica.* 47, 5–6.

Sasidharan, K.R., B. Nagarajan, V. Mohan, A. Nicodemus, N.P. Mahadevan, A. Durai, and K. Gireesan (2003). Insect pollination versus enhanced fruit production in *Tamarindus indica* and *Tectona grandis* (Pollination Ecology Volume). *Journal of Palynology.* Today and Tomorrow's Printers & Publishers, New Delhi, India. 35/36: 93-97.

Thimmaraju, K.R., M.A. Reddy, N. Swamy, and U.V. Sulladmath (1977). Studies on the floral biology of tamarind (*Tamarindus indica* L.). *Mysore Journal of Agricultural Science*, 11: 293-298.

Usha, K. and Singh B. (1994). Selection of tamarind tree types for higher yield. *Crop Improvement*, 23(2): 199-202.

Worthington, T.B. (1959). Ceylon Trees. The Colombo Apothecaries Co., Colombo: 429.

■■■

Chapter 8

GENETIC IMPROVEMENT OF *TAMARINDUS INDICA* L.

Dr. A. Mayavel, Dr. B. Nagarajan, Dr. A. Nicodemus & Mr. R.S. Prasanth

Institute of Forest Genetics and Tree Breeding, Coimbatore, Tamil Nadu, India

Tamarindus indica L. (Tamarind) is found throughout much of the tropics. It is grown unattended in backyards, roadsides or wastelands (Gunasena and Hughes, 2000). It is suited for avenue plantings in roadsides, in and around villages as multipurpose trees for agroforestry systems.

Conservation of tamarind germplasm *in situ* and *ex situ* is an important prerequisite to prevent loss of genetic diversity of the species (Gill, 1984). The information on research on the ecology and value of tamarind is scanty. Before germplasm can be collected and tree improvement programmes initiated, there is a need to study the ecology of tamarind trees in its natural habitats and to determine the *in situ* and *ex situ* conservation strategies. Furthermore there is a need to develop guidelines for appropriate interventions for conservation and increase utility of the tree (Gunasena, 1997)

To meet the food security of exploding human population, there is a need to identify and use tree resources which could offer the food security and income to the rural populations. Tamarind improvement programme in relation with conservation and utilization have been restricted in India. It is noted that the germplasm is being eroded in India and *in situ* and *ex situ* conservation and management is required. The major limitation for domestication of the underutilized species is lack of research for its improvement at local, regional and international levels (Meghwal, 1997) and improved germplasm are not freely available in many countries (Sanchez, 1984)

Genetic erosion is occurring in most of the growing areas due to deforestation resulting from rural development and urbanisation. Population pressure on restricted ecosystems and damaging shifts in land use patterns with increasing areas being planted with perennial cash crops are main recent affecting tamarind germplasm conservation. Selective felling of individual trees for timber and fuel wood production and inadequacy of local and international markets for the fruit pulp is also causing genetic erosion and negligence of tamarind. The consequences are the loss of genetic resources, which may threaten its own survival and its possible use in the future. Therefore, a systematic germplasm maintenance and conservation programme for tamarind will be beneficial both in the short and long term for all the growing countries. It is only through an organised strategy that germplasm can be supplied continuously for tree improvement in the future.

Genetic improvement through the use of superior clones has been described by Kulkarni *et al.* (1993). Tamarind has a relatively long generation time and is primarily outcrossing, thus any conventional breeding approaches would require considerable investment in time and money. There are trees with very high yielding potential exceeding 800 kg/tree/year and such trees could be selected for vegetative propagation by air layering or grafting methods to produce fast growing trees for local (home garden) and commercial (orchard) use. Such trees can be grown with comparative ease and minimum management and should prove to be profitable due to their commercial value for small-scale farmers in African and Asian countries.

Germplasm must to be rigorously characterised and evaluated to select desirable traits. If fast growing and high yielding strains are selected for different uses, tamarind would rapidly become a desirable tree for small scale plantations in a short period, particularly since it continues to bear fruit for many years. More - trait specific research and combining desirable characters together is needed to develop cultivars. Equal emphasis should be given to select clones for Agroforestry systems and small scale plantations. The desirable ideotypes should be developed to fit into various niches as well as for more intensive commercial production. This will require more attention to tissue culture since it offers a way of cleaning stocks from diseases.

8.1. SELECTION

The selection of elite trees is an important step for any tree improvement programme. Tamarind trees selected based on important characteristics such as acidity of the pulp, content of tartaric acid and sugar, real value of pulp, pod bearing ability (flowering and fruit maturing), pod size, pulp, fibre and seed weights, and number of seeds. In India, local farmers usually identify mother trees that consistently produce large numbers of fruits for their seed collection and propagation. Some of the trees that have been selected on these bases are reported to have larger pods measuring of 25 cm long and 5 cm wide (Jambulingam and Fernandes, 1986). Selections have been made from natural stands growing in these countries In Thailand, more than 50 sweet tamarind cultivars are grown, while in the Philippines, eight selected cultivars are popular among the farmers. These sweet types have created a resurgence of interest in many countries of the Southeast Asian region and encouraged researchers to undertake studies on tamarind.

The sweet tamarind has been attributed to a point mutation. Occasionally isolated branches on a tree may bear sweet fruits while others bear normal sour ones. Bud sports have been propagated vegetatively and form the basis for a range of recent cultivars

Although tamarind is an ancient domesticate, little attempt has been directed to its genetic improvement. This is understandable because tree improvement research that combines developmental and operational phases is time consuming and the large scale cultivation of tamarind is still in its early stages. Indigenous farmers have however selected planting materials from natural populations based on desirable and observable characteristics but such phenotypic selection means the growing stocks are virtually wild. Tamarind was recorded over a century ago as a variable species especially for pulp colour and sweetness. Since these species exhibits extensive variation in characters such as foliage, flower and pod production and timber quality. There is a considerable scope to improve the species. Improvement holds the key for boosting productivity and yield of the orchards and involves development of genotypes possessing desirable characters like fast growth, good tree form, high yield and resistance or tolerance to major pests, diseases and drought (Radhamani *et al.*, 1998). Since the variation in pod length and pod width was found to be genotypic (Shivanandam and Raju, 1988 and Hanamashetti, 1996). Similarly for other trials the potential for improvement depends on sampling the genetic variability available within and between populations. Hence, knowledge of genetic variation and structure of a species and genetic parameters of important traits are essential to develop effective improvement and conservation strategies (Table 8.1).

Table 8.1. Variation in Vegetative and Reproductive Characters of Tamarind

S.No	Characters	Variation
1	Terminal shoot length	Short-long
2	Flowering behaviour	Early-mid-late
3	Number of primary branches	5-8
4	Number of secondary branches	30-63
5	Number of tertiary branches	45-91
6	Number of inflorescences/branch	531-990
7	Number of flowers/branch	5697-15587
9	Pod length (cm)	4-30
10	Pod width (cm)	1.9-4.0
11	Pod weight (g)	4.5-60.0
12	Amount of pulp in ripened fruit (%)	30-50

S.No	Characters	Variation
13	Single seed weight (g)	0.8-1.2
14	Number of seeds/pod	4-10
15	Seed weight (g)	1.6-10.0
16	Sweetness of pulp	Sweet-sour
17	Fibre content	Low-high
18	Pod yield per plant (kg)	Very variable

(Challapilli, 1995; Divakara, 2002; Shanthi, 2003)

8.2. EXPLOITING VARIATION

Wide variation was observed on phenology, flowering and fruiting characters in within and between population which will helps breeder to select superior candidate plus trees and multiply same in large scale. Higher variation helps the breeder to combine desirable characters for qualitative and quantative improvement. The variation on vegetative and reproductive characters were studied in different population and described.

Based on the assessment of Candidate Plus Trees and their progenies Shanthi (2003) concluded that pod size, tree height, number of flowers, number of inflorescences and number of tertiary branches could be used as measures for evaluating tamarind trees in order to select superior trees for further evaluation. The experiments carried out in Coimbatore revealed that the highest phenotypic variation occurred for pod length and the lowest for seed weight whilst the maximum genotypic variation has been observed for pulp weight. The highest phenotypic and genotypic coefficients of variation have been recorded for pulp weight (Chundawat, 1990; Challapilli *et al.*, 1995; Divakara, 2002; Shanthi, 2003). A maximum heritability of 0.5 and the highest genetic advance percentage were recorded over a mean of 42.5 for pulp weight. The highest genetic advance was observed for pulp weight. Pod length showed highest positive phenotypic correlation with pulp weight. Hence, selection based on pulp weight and pod length is useful in tamarind improvement programmes. Further, because of the presence of significant positive correlation coefficients between pod length and pod width, they are suggested as selection criteria for the identification of superior trees.

Institute of Forest Genetics and Tree Breeding have initiated both short and long term improvement programmes for *Tamarindus indica.* A comprehensive approach that attempts to study the variability and biology of the species is followed by the institute. IFGTB's various activities on Tamarind improvement are given below.

Selection of superior candidate plus trees is an important step for any tree improvement programme. In India farmers selecting the tamarind trees which are consistency produce higher yield over a period time for collection and propagation. Institute of Forest and tree breeding initiated systematic short and long term tree improvement programme of tamarind and established progeny and clonal trial.

Selection criteria for selection of superior CPT's of *Tamarindus indica*

1. The tree should posses round large canopy with many branches and good health condition
2. The terminal shoot of tamarind trees should be larger
3. The tree should have early, longer flowering period and regular bearing habit.
4. The fruit should be heavy, fleshy, acidic and high pulp recovery.
5. The tree should have higher yield.
6. The tamarind trees should have wider adaptability and resistance to pest and disease.

A survey was conducted in 1994 to identify the phenotypic variability in the Kanyakumari district of Tamil Nadu. Thirteen phenotypes were identified with distinct fruit and pulp characteristics. Fruit length, breadth and weight, pulp weight, seed length, breadth and weight, tree height, and girth were the characteristics indicated for different phenotypic selection. The pulp weight is found to be positively correlated with fruit length, breadth and weight and also seed weight.

8.3. PROGENY TESTING

Open pollinated seeds were collected from selected phenotypes and a progeny trial has been laid out with 13 half sib families in a completely randomized design at IFGTB Campus, Coimbatore. Data on early growth and branching characteristics are being recorded. The trial will help to evaluate the heritability of improvement characteristics and variability among the phenotypes.

8.4. SELECTION OF RARE TAMARIND GENETIC RESOURCES

Intensive survey was carried out in different parts of Tamil Nadu, Karnataka, Andhra Pradesh and Puducherry to locate the trees. To know the availability of red and sweet tamarind communication was made with forest officials, farmer's forum, Agricultural and Horticultural officers. News paper advertisement in leading news papers and farmers oriented magazine in vernacular languages was given and about 36 red tamarind and 30 sweet tamarind trees were identified in different parts of southern India.

8.5. GERMPLASM BANK

All the identified tamarind trees were vegetative multiplied through cleft grafting. The conservation of rare tamarind genetic resources was done by establishing Germplasm bank of red and sweet tamarind at Kurumpapatti, Salem (RF). Also the best genetic resources selected by Tamil Nadu, Karnataka and Andhra Pradesh Forest Department are assembled at tamarind germplasm bank. At present Institute of forest genetics and tree breeding maintaining 100 tamarind clones selected from different parts of country. These Germplasm will act as basic genetic resources for future tree improvement programs (Table 8.2).

Table 8.2. Details of phenotypes of *Tamarindus indica* selected by IFGTB

Sl. No	Pod characteristics	*Location
1	Straight and flattened with sweet taste	Pallur
2	Straight, broad and fleshy	Pallur
3	Straight fleshy	Nattalam
4	Straight and broad	Annan villai
5	Straight and flattened	Thiruppara malai
6	Straight with one constriction	Nattalam
7	Straight with three constrictions	Kara villai
8	Straight, rounded and many constrictions	Kara villai
9	Straight and curved at base	Kara villai
10	Straight curved and flattened	Kalluvittan villai
11	Straight curved and rounded	Nattalam
12	Curved and flattened	Kara villai
13	Curved with less flesh	Karungal

8.6. HYBRIDIZATION FOR QUALITATIVE IMPROVEMENT

Tamarind hybrids are produced by control crossing two different tamarind clones. Improving qualitative and quantitative value of tamarind clones is possible through hybridization. The Institute of Forest Genetics and Tree Breeding initiated hybridization programme for qualitative improvement of high yielding tamarind clone. The main objective is to combine red colour (Anthocyanine) from red tamarind in to high yielding tamarind clones. The full diallele method was followed for executing control hybridization (Zobel and Talbert, 1984). About nine high yielding clones of tamarind were selected from clonal assemblages at Theni and Vaigai dam Flowers were emasculated using a clean fine tip forceps (from 15.00 hrs to 20.00 hrs) and red tamarind pollen were collected from red Jayamangalam clone. Flowers were dusted with red tamarind pollen with a dry paint brush or needle from 6.00 hrs to 11.00 hrs on the next day. In apomictic treatments the emasculated flowers were not dusted with pollen. In an inflorescence only the treated flower were retained. In each clone 100 flowers were operated per day per treatment and all treatments were repeated for seven days. Operated flowers were caged in paper covers (size of 12 cm x 7 cm) and tagged properly. Bags were removed on the seventh day for recording fruit setting. About 180 days after crossing fruits were harvested and seed extracted. Hybrid trails were established at State Forest Research Institute, Kolapakkam and Forest campus, Coimbatore, Tamil Nadu. The hybrid seedlings were planted in 10 x 10 m spacing in randomized block design with four replications. The hybrid progenies initiated flowering and fruiting 6 years after planting. Among different combination of hybrid progenies Vembrampattu 416 (VEP 416) x Red Jayamangalam (RJ 403) found to be superior in qualitative and quantitative traits

8.7. CLONAL SEED ORCHARD

Tamil Nadu Forest Department conducted extensive field survey to identify tamarind trees with desirable quantative and qualitative characters. About 40 Candidate Plus trees were identified and multiplied clonally. Tamarind clonal seed orchards were established at Neyveli (Cuddalore), Vringipuram (Vellore) Kolapakkam (Kancheepuram) and Theni (Theni). The list of clones assembled in clonal orchard are given in Table 8.3.

Table 8.3. Vegetative and reproductive characters of Tamarind clones

Sl.No	Clone Name	Bearing Habit	Flowering Season	Yield (kg.)
1	Kandukarapalli	Alternate	Early	48
2	Red pulp	Regular	Early	35
3	Periyakulam 1	Regular	Mid	120
4	Pollachi 11	Irregular	Late	52
5	Salem 226	Alternate	Mid	46
6	Thirupathur 145	Regular	Mid	33
7	Thirupathur 142	Regular	Mid	41
8	Denikanikottai 112	Regular	Early	53
9	Hassanur 1	Alternate	Early	65
10	Vellore 67	Regular	Late	72
11	Reddyur 141	Alternate	Mid	30
12	Denikanikottai 106	Regular	Early	48
13	Neyveli 49	Alternate	Mid	23
14	Pallukapalli APP	Irregular	Mid	27
15	Chengam 143	Alternate	Mid	50
16	Chittur 163	Alternate	Late	42
17	Chittur 166	Alternate	Late	46
18	Chittur 162	Alternate	Late	48
19	Vellore 40	Regular	Late	60

Sl.No	Clone Name	Bearing Habit	Flowering Season	Yield (kg.)
20	Chittur 165	Regular	Late	41
21	Chittur 164	Regular	Late	43
22	Vellore 58	Alternate	Late	62
23	Vembara patti 412	Regular	Mid	73
24	Hassanur 3	Alternate	Early	60
25	Vellore 57	Regular	Late	63
26	Chittur 161	Alternate	Late	45
27	Hassanur 4	Alternate	Mid	58
28	Hassanur 161	Regular	Mid	53
29	Vellore 64	Regular	Late	60
30	Vellore 1	Regular	Early	130
31	Seedless	Irregular	Early	15
32	Hassanur 10	Regular	Mid	43
33	Neyveli 10	Regular	Mid	31
34	Denikanikottai 161	Regular	Early	85
35	Vellore 66	Alternate	Late	67
36	Vellore 2	Regular	Late	65
37	Vellore 3	Alternate	Late	60
38	Hassanur 2	Regular	Mid	82
39	Devarulimangalam	Regular	Early	140
40	Vellore 68	Regular	Late	75
41	Thirupathur 50	Regular	Mid	60
42	Vellore 59	Regular	Late	69

8.8. MOLECULAR MARKER STUDIES

Molecular and bio chemical markers can be used to estimates the genetic variability in tamarind genotypes. In IFGTB two enzyme systems namely peroxidases and esterases have been used to differentiate thirteen phenotypes. Preliminary results have revealed that the different phenotypes selected do not vary much genetically. Both parents and progenies were found to have the same isozyme banding patterns for both the loci. Further, isozyme studies have been undertaken to study the variability and breeding system in 25 populations of Tamarind in Tamil Nadu.

Nandini *et al.* (2011) observed RAPD profiles of different individuals were compared and a total of sixty one scorable bands were produced with six primers ranging from eight for OPA-R15 to thirteen for OPA-Y01. The dendrogram showing clustering of tamarind genotypes based on morphological studies were not in accordance with molecular markers. Genotypes which were closely related were found to be unrelated at the molecular level. A sizeable amount of intra population diversity recorded in the present study can be used to identify diverse parents which can be utilized in hybridization programmes to efficiently introgress the desirable trait of interest.

8.9. FUTURE PROSPECTS

- Systematic Germplasm collection from centre of origin and other tamarind growing areas.
- Selection of tamarind trees with desirable characters like higher fruit yield, regular bearing habit and resistance to pest and disease
- Introduction of high yielding sweet tamarind from Thailand and Philippines
- Standardization of precision silviculture techniques for enhancing flowering and fruiting.
- Detailed investigation of taxonomic status of red and sweet tamarind
- Qualitative and quantative improvement through control hybridization and testing of hybrids
- Developing descriptors for tamarind germplasm
- Systematic tree improvement/breeding program with multi location testing of superior clones
- Identification of tamarind clones suited for degraded soil and waste land for rural employment generation and poverty alleviation

REFERENCES

Challapilli, A.P., Chimmad, V.P. and Hulamini, N.C. (1995) Studies on correlation of some fruit characters in tamarind fruits. *Karnataka Journal of Agricultural Science,* **8(1)**: 114-115.

Chundawat, B.S. (1990): Arid Fruit Culture, New Delhi, India. Oxford & IBH Publication Co. Pvt. Ltd.

Divakara, B.N. (2002). Clonal evaluation and genetic diversity studies using biometric and iso enzyme approaches in *Tamarindus indica* L. *Ph.D. (For) Thesis.* Tamil Nadu Agricultural University, Coimbatore.

Gill, K.S. (1984) Research imperative beyond the green revolution in the third world. In Human fertility, Health and Food Impact on Molecular Biology and Technology.

Gunasena, H.P.M. (1997) Domestication of agroforestry trees in Sri Lanka. Proceedings of the Regional Workshop on Domestication of Agroforestry Trees in Southeast Asia. A publication of Win rock International in collaboration with ICRAF. Yogyakarta, Indonesia pp. 49-53.

Gunasena, H.P.M and Hughes, A. (2000). Fruits for the future – Tamarind (*Tamarindus indica* L.). International centre for underutilized crops, Southampton, UK, pp.170.

Hanamashetti, S.I. (1996). Vegetative propagation and genetic diversity evaluation in Tamarind. *Ph.D. (Hort.) Thesis*, UAS, Dharwad. Karnataka.

Meghwal, P. R. (1997) Exploiting under-utilised fruits of the arid zone. *Indian Horticulture*, October-December: **26-27**.

Nandini, Ramesh, Telem Ratan Singh and G.N. Dhanapal (2011). Morphometric and molecular diversity studies in Tamarind (*Tamarindus indica* L.) population. *J. Plant Mol. Biolo. Biotechnology,* **2(2)**: 26-33.

Radhamani, A., A. Nicodemus, B. Nagarajan and A.K. Mandal (1998). Reproductive Biology of Tropical Tree species. In: Forest Genetics and Tree Breeding, A.K. Mandal and G.I. Gibson (Eds) CBS Publishers, New Delhi.

Sanchez, P. C. (1984) Tropical fruit wines. A lucrative business. Research at Los Baños, The University of the Philippines 3: 10-13 (FSTS: 86-05H0063).

Shanthi A (2003). Studies on variations and associations in selected populations, plantations and clones in tamarind (*Tamarindus indica* L.). Un published *PhD Thesis.* Bharathiar University, Coimbatore, India.

Shivanandam, V.N. and Raju (1988). Studies on systematic selection of production types of tamarind (*Tamarindus indica* L.) types. *Mysore J. Agril. Sci.,* **22(2)**: 229-231.

Zobel, B., and J. Talbert (1984). Applied Forest Tree Improvement. John Wiley and Sons New York, USA.

■■■■

CHAPTER **9**

PRODUCTION, MARKETING AND VALUE CHAIN OF TAMARIND IN INDIA

Dr. M. Chinnadurai, Dr. S. Varadha Raj & Dr. R. Chitra

Tamil Nadu Agricultural University, Coimbatore

Tamarind (*Tamarindus indica*; Fabaceae) is probably indigenous to tropical Africa and it has been cultivated in India. It is being cultivated from trees growing all types of lands such as forest, common and private. It is widely distributed throughout the tropical belt. India is the largest producer of Tamarind in the world. The major areas of production are in Cameroon, Central African Republic, Ethiopia, Guinea, Kenya, Nigeria, Senegal, Sudan, Tanzania, Uganda, Afghanistan, Bangladesh, Brazil, Cambodia, China, Colombia, Cuba, Egypt, Malaysia, Mexico, Myanmar, Nicaragua, Pakistan, Philippine, Srilanka, India, and Thailand (Singh *et al*, 2007). The Sweet Tamarind is harvested at ripened stage and usually consumed as fresh, while the sour Tamarind is usually processed into a range of value added products (Joshua, 2006).

9.1. INDIAN SCENARIO OF TAMARIND

India is the world largest producer of Tamarind. Madhya Pradesh, Bihar, Andra Pradesh, Karnataka, Tamil Nadu and West Bengal are the leading states in production of Tamarind in India. The area and production of the Tamarind was 29831 ha and 82901 tonnes respectively in 1992 and increased to 59000 ha and 188000 tonnes respectively in 2014, whereas the productivity was 2.78 tonnes/ha in 1992 and 3.74 tonnes /ha in 2014 (Fig. 9.1) (Hand Book on Horticulture Statistics. 2014).

Andhra Pradesh produces about 70,000 quintals of Tamarind, out of which the bulk comes from the farm sector. The Scheduled Tribes collect only about 40,000 quintals. It is estimated that around 1.10 lakh families are involved in the retail sale of Tamarind through the Girijan Cooperative Corporation (GCC) distribution network and auction cum tender process (FGLG Report, 2008). Among states, Tamarind is largely produced and consumed in Tamil Nadu. The period of collection of Tamarind fruits is February – April and sometimes extend to May. The best fruiting occurs on alternate years. Tamarind is collected from trees growing on forest, common, and private lands. Tamarind has a huge domestic and overseas market and commands a good price from the end consumers.

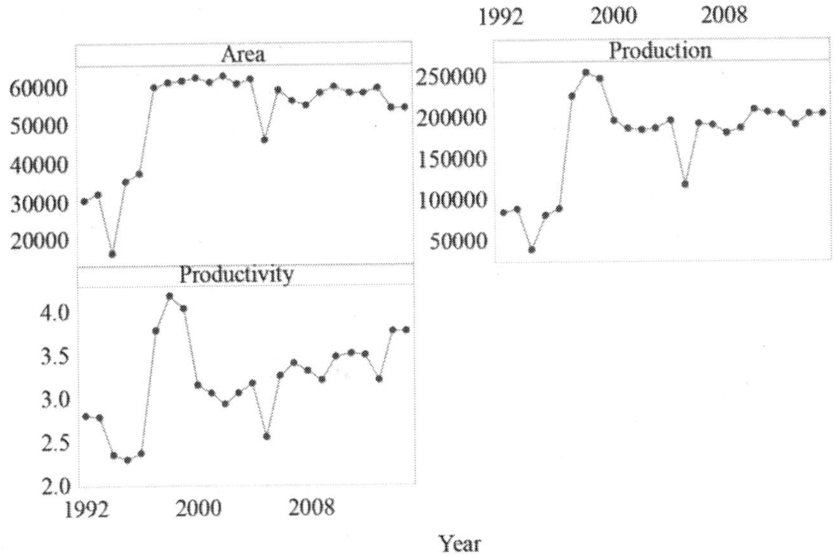

Fig. 9.1. Area, Production and Productivity of Tamarind in India (2007-2015) (Source: www.indiastat.com; Hand Book on Horticulture Statistics, 2014)

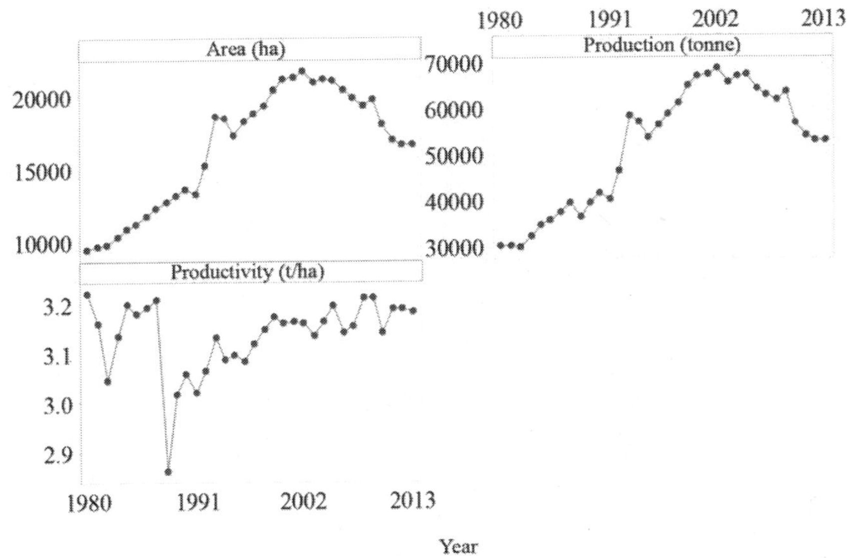

Fig. 9.2. Area, Production and Productivity of Tamarind in Tamil Nadu (1980-2013)

In Tamil Nadu, the area under Tamarind has increased from 9393 ha in 1980-81 to 16640 ha in 2013-14 is given in Fig.9.2. There was a shift in the area under Tamarind during 2000-05. After 2006-07, there is a further downward shift in the area under Tamarind might be due to climatic factors and other social factors. The production of Tamarind in the above period shows a significant increase. The production of Tamarind has increased from 30270 tonnes (MT) in 1980-81 to 53000 tonnes in 2013-14. It is considered that the yield per hectare decreased from 3.22 tonnes/ha in 1980-81 to 3.19 tonnes / ha in 2013-14. This declined productivity may be due to the age old plantation and lack of cultural operation. Uneconomic plantations have to be replaced with high yielding varieties, use of fertilizer, assured irrigation and timely intercultural operation to maximize the productivity of Tamarind in Tamil Nadu.

9.2. GROWTH RATE OF TAMARIND AREA, PRODUCTION AND PRODUCTIVITY

The annual compound growth rate of Tamarind area, production and yield in India are depicted in Fig. 9.3. The annual CGR of Tamarind area was 2.75 per cent, 3.95 per cent, 1.18 per cent respectively during 1992-2014. Based on CGR, Tamarind area, production and productivity will be 63738 ha, 239530 tonnes and 3.77 t/ha respectively in 2020 and it will be 72 981 ha, 290749 tonnes and 4.00 t/ha respectively in 2020 (Table 9.1).

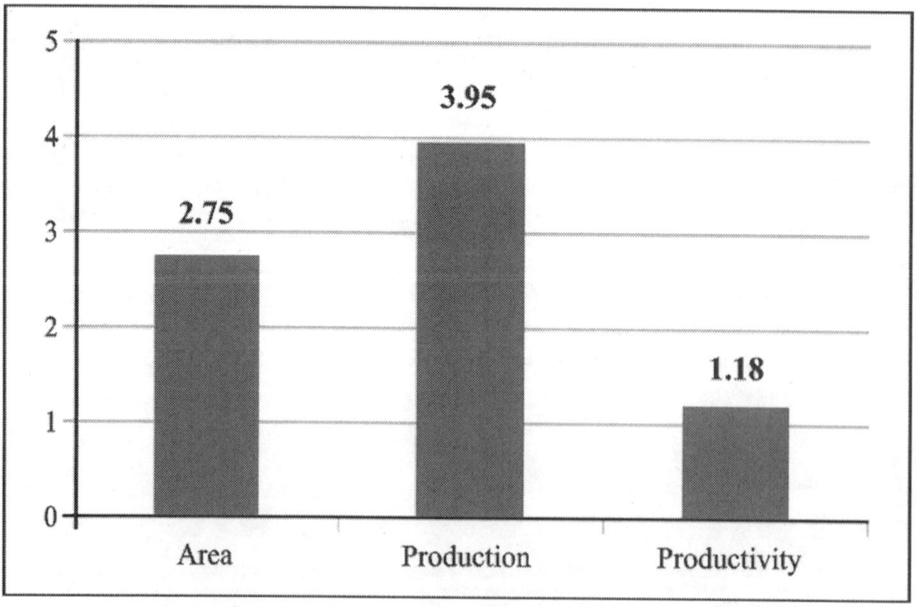

Fig. 9.3. GCR of Area, Production and Productivity of Tamarind in India (1992-2014) (Source: www.indiastat.com)

Table 9.1. Forecasted Area, Production and Productivity of Tamarind in India

Year	Area (ha)	Production (tonnes)	Productivity (t/ha)
2016	57195	205131	3.60
2017	58765	213238	3.64
2018	60378	221664	3.68
2019	62036	230424	3.73
2020	63738	239530	3.77
2021	65488	248995	3.82
2022	67286	258835	3.86
2023	69133	269063	3.91
2024	71031	279696	3.95
2025	72981	290749	4.00

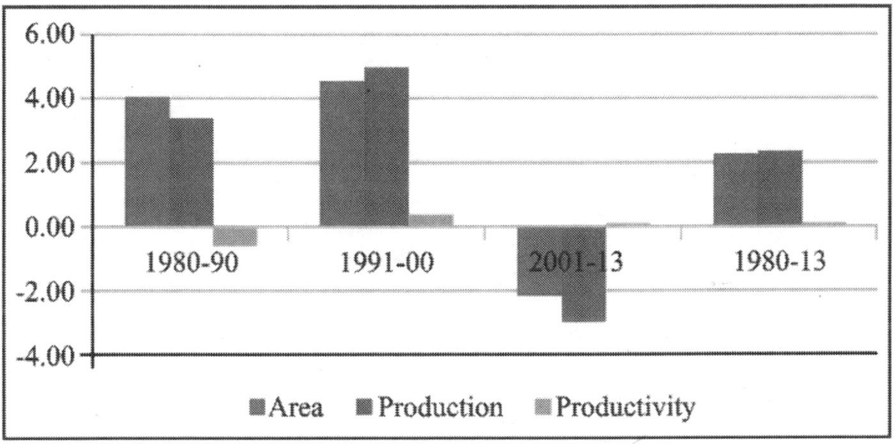

Fig 9.4. Growth rate in Area, Production and Productivity of Tamarind in TN (Source: www.indiastat.com)

In Tamil Nadu, the annual CGR of Tamarind area was 4.02 per cent, 4.56 per cent, -2.15 per cent and 2.27 per cent in 1980-90. 1991- 2000, 2001-13 and 1980-2013 respectively (Fig. 9.4). The area of the Tamarind was reduced at the rate of 2.15 per cent during 2001-13. In the long run, there is a positive growth rate (2.27 per cent) during 1980-13.

The annual CGR of Tamarind production was increased from 3.40 per cent in 1980-90 to 4.98 per cent in 1991-2000. There was a negative growth rate of production (-2.97) was realized during 2001-13. The significant decrease in the growth rate of production was mainly due to stagnation of the area and coupled with decreased productivity. In the long run (1980-13), the growth rate of the Tamarind production was 2.35 per cent in Tamil Nadu. The yield growth rate of the Tamarind was -0.60 per cent during 1980-90 and it was 0.08 per cent for the period of 1980-13. There is no remarkable change in the growth rate in the short run and long run. This is due to lack of high yielding and short rotation Tamarind crops in Tamil Nadu.

9.3. DISTRICT WISE SCENARIO OF TAMARIND

The district wise area, production and productivity of Tamarind are depicted in Fig 9.5-9.7 and Table 9.2.

Table 9.2. Districtwise Area Growth Rates of Tamarind in Tamil Nadu

Positive		Negative	
District	**CGR (%)**	**District**	**CGR (%)**
Nagapattinam	4.55	Thoothukudi	-0.40
Virudhunagar	1.07	Sivagangai	-0.87
Ariyalur	0.14	Tirunelveli	-0.94
Karur	0.01	Madurai	-1.16
		Thanjavur	-1.38
		Krishnagiri	-1.46
		Tiruvarur	-1.55
		Dindugal	-1.71
		Tiruchirapalli	-2.00
		Theni	-2.22
		Cuddalore	-2.73
		Kancheepuram	-3.22
		Vellore	-3.22
		Pudukottai	-3.35
		Villupuram	-3.83
		Ramanathapuram	-4.17
		Kanyakumari	-4.99
		Namakkal	-5.65
		Salem	-6.01
		Erode	-6.09
		Coimbatore	-6.24
		Tiruvannamalai	-9.37
		Dharmapuri	-9.95
		Tiruppur	-13.39
		Permbalur	-15.38
		Thiruvallur	-16.23

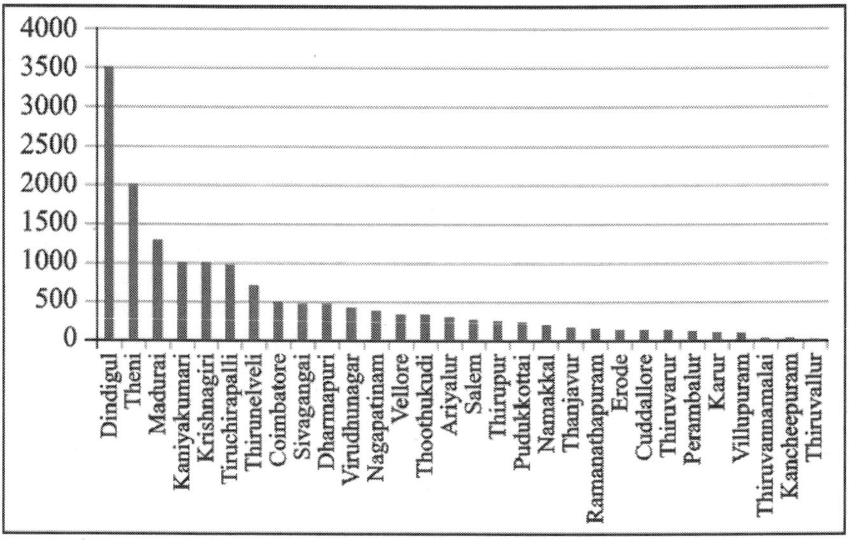

Fig. 9.5. Area of Tamarind in Tamil Nadu during 2013-14
(Source: www.indiastat.com and season and crop report, 2014)

Out of 15850 ha in Tamil Nadu, the highest area was found in Dindigul district (3498 ha), followed by district of Theni (1283 ha), Kanyakumari (988 ha), Krishnagiri (978 ha), Trichirappalli (939 ha) and Tirunelvali (709 ha). The positive growth rate was found in Nagapattinam (4.55%), Virudhanagar (1.07%), Ariyalur (0.14%) and Karur (0.01%) and other remaining districts had a negative growth rate, out of which, the highest decline of growth rate was found in the district of Thiruvallur (-16.23%), Permbalur (-15.38%), Tiruppur (-13.39%) Dharmapuri (-9.95%) and Tiruvannamalai (-9.37%) (Fig. 9.5).

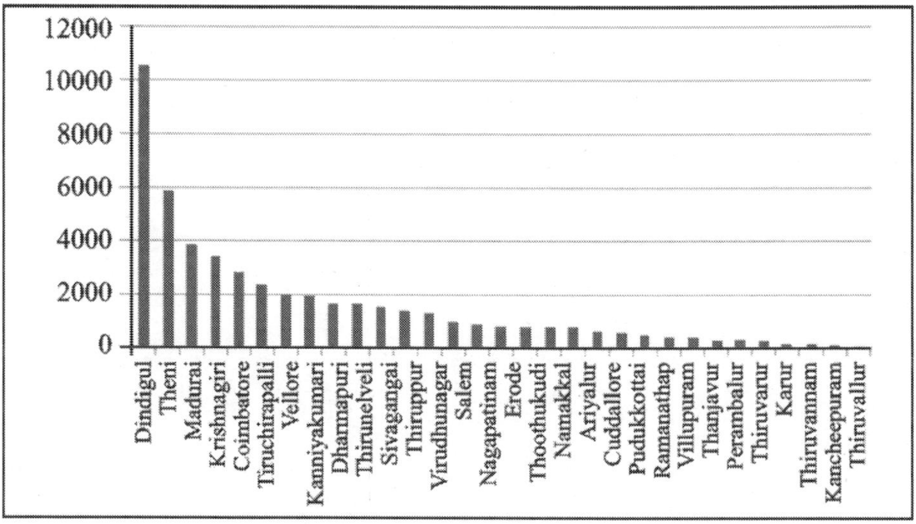

Fig. 9.6. Production of Tamarind in Tamil Nadu during 2013-14

The highest Tamarind producing districts were Dindigul (10543 t), Theni (5880 t), Madurai (3872 t), Krishnagiri (3407 t), Coimbatore (2820 t), Tiruchirapalli (2345 t) and Vellore (1984 t) in 2013-14 (Fig. 9.6).

Table 9.3. Districtwise Production Growth Rates of Tamarind in Tamil Nadu

Positive		Negative	
District	**CGR (%)**	**District**	**CGR (%)**
Nagapatinam	4.74	Thoothukudi	-0.33
Dindigul	2.93	Thirunelveli	-0.73
Virudhunagar	1.07	Sivagangai	-1.05
Ariyalur	0.59	Madurai	-1.16
Karur	0.01	Thanjavur	-1.39
		Thiruvarur	-1.53
		Krishnagiri	-1.53
		Tiruchirapalli	-2.01
		Theni	-2.16
		Cuddallore	-2.69
		Vellore	-3.22
		Pudukkottai	-3.38
		Kancheepuram	-3.74
		Villupuram	-3.78
		Ramanathapuram	-4.16
		Erode	-4.80
		Kanniyakumari	-5.00
		Namakkal	-5.32
		Salem	-5.86
		Coimbatore	-5.91
		Thiruvannamalai	-8.62
		Dharmapuri	-9.96
		Perambalur	-14.71
		Thiruvallur	-16.05
		Thiruppur	-23.48

It could be seen from the table that, except Nagapatinam (4.74%), Dindigul (2.93%), Virudhunagar (1.07%), Ariyalur (0.59%) and Karur (0.01%) districts were registered a negative CGR of production. The highest negative growth rate of production was noted in Thiruppur (-23.48%), Thiruvallur (-16.05%),

Perambalur (-14.71%), Dharmapuri (-9.96%), Thiruvannamalai (-8.62%), Coimbatore (-5.91%) and Salem (-5.86%) (Table 9.3).

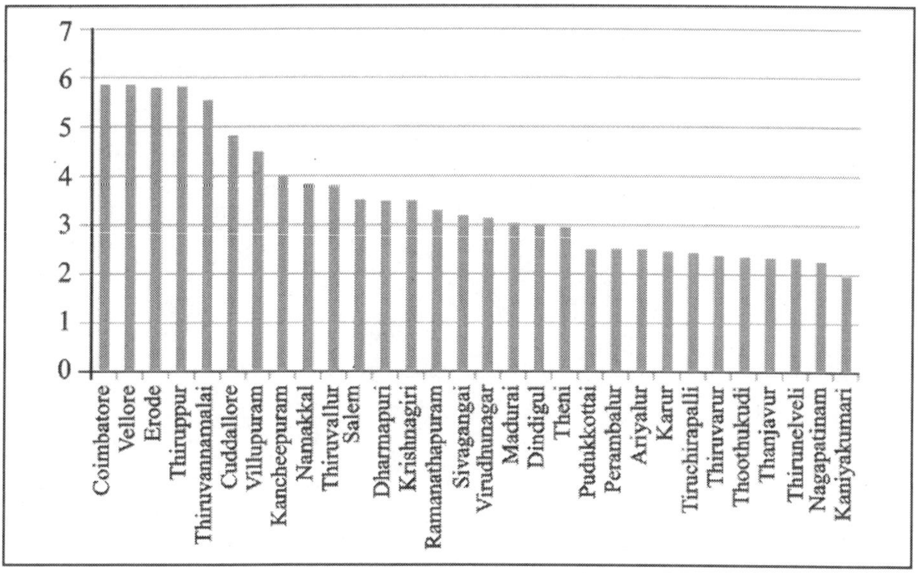

Fig. 9.7. Productivity of Tamarind in Tamil Nadu during 2013-14

Tamarind trees are available in large numbers all over Tamil Nadu, especially along road avenues, patta fields, revenue lands, village common lands, Temple lands etc. Their production is fairly good and is partly meeting the current demands. The highest productivity was noticed in the district of Coimbatore (5.86 t/ha), Vellore (5.84 t/ha), Erode (5.81 t/ha), Thiruppur (5.8 t/ha), Thiruvannamalai (5.54 t/ha), Cuddallore (4.83t/ha), Villupuram (4.5 t/ha), Kancheepuram (4 t/ha) and Namakkal (3.81 t/ha). The positive growth rate of productivity was found in the district of Dindigul (4.72%), Erode (1.36%), Thiruvannamalai (0.94%), Perambalur (0.81%), Namakkal (0.35%) and Coimbatore (0.35%). The highest negative growth rate of productivity was found in Thiruppur (-7.26%), Thanjavur (-1.55%) and Ariyalur (-0.68%) district.

According to the TASPEF report (20011-12), the average yield of the Tamarind fruit (dried) in the conventional forest area was 0.806 kg per tree, whereas it was 40 kg per tree in Patta lands in Javadhis tribal areas which generated nearly Rs.4.00 crore from 34088 trees in 2011-12. The Tamarind trees in Patta yield 40 to 50 times higher than that of forest trees. Hundred kilograms of raw Tamarind will yield 30 kg of pod removed and de-seeded Tamarind. Large scale Tamarind plantations are available in Tirupattur, Harur, Dharmapuri, Hosur, Salem, Namakkal and Dindigul divisions. The study indicated that full potential was not being tapped and also entire collection was not properly accounted in the natural forest area. In general, the potential yield of forest Tamarind trees is nearly 10 kg per tree. The reasons for very poor Tamarind collection is due to difficulties in accessing remote areas, scattered trees, wildlife damages, improper harvesting methods and non remunerative price offered by traders and lack of adequate awareness and marketing skills of Village Forest Council (VFC) members which delayed their decision making process in selling their produce by the mean time, the matured fruits may fall down and decay (Table 9.4).

Table 9.4. Districtwise Productivity Growth Rates of Tamarind in Tamil Nadu

Positive		Negative	
District	**CGR (%)**	**District**	**CGR (%)**
Dindigul	4.72	Thiruvarur	0.00
Erode	1.36	Cuddallore	0.00
Thiruvannamalai	0.94	Madurai	-0.01
Perambalur	0.81	Ramanathapuram	-0.01
Namakkal	0.35	Dharmapuri	-0.01
Coimbatore	0.35	Tiruchirapalli	-0.03
Thirunelveli	0.20	Pudukkottai	-0.04
Nagapatinam	0.17	Krishnagiri	-0.07
Salem	0.11	Sivagangai	-0.19
Thoothukudi	0.07	Ariyalur	-0.68
Theni	0.07	Thanjavur	-1.55
Virudhunagar	0.03	Thiruppur	-7.26
Kancheepuram	0.02		
Thiruvallur	0.02		
Villupuram	0.01		
Kanniyakumari	0.01		
Karur	0.00		
Vellore	0.00		

9.4. METHOD OF HARVEST AND MARKETING OF TAMARIND

Tamarind plays a significant role in meeting the socio-economic and livelihood needs of rural communities. In accordance with the national forest policy 1988, the government of Tamil Nadu initially gave the free NTFP benefits to VFCs formed under Tamil Nadu Afforestation Project (TAP) scheme up to 1977 and then, it was extended to tribal and other forest dwelling communities since 2003. In Tamil Nadu, there are 17 major NTFP species predominantly contributing to the economy. Out of which, Tamarind is considered as an important non-timber forest produce in view of their production, yield, utility and benefits to the dependent communities. TAP through Joint Forest Management (JFM) concept and formation of VFCs to involve local people in tree growing activities on degraded forest lands. The Government also extended the processes of benefit sharing from the forest areas lying within their management units. The forest produces were given free of cost to the VFC members in return for their support and cooperation for the management of the forest allotted to the VFC. The benefits were also extended to Joint Forest Management Committees formed under National Afforestation Programmes (NAP) implemented through Forest Development Agencies (FDA). In both TAP and NAP projects, all community people residing in programme villages and members of VFCs are getting benefits.

In forest department area, the area is assessed by its potential trees per unit area. The value of the forest area is fixed and invited a closed tender from registered traders. The floor price is fixed by the forest department. The area is declared for auction and then the highest bidder will be given a chance to collect Tamarind fruits in the prescribed forest area. The VFC will give the area to the highest bidder of private contract. In accordance with permission of Divisions District Forest Officers, their contracts will be finalized and arrange to deposit the sale amount into the village development fund of the VFC. In some area, VFC collects Tamarind and sell to private traders. In Tamil Nadu, Tamarind fruits are collected in different stages. The ripened fruits will start to fall in the month of January and that will be collected from the ground. It is an indication of fruit maturity. The mature tree's all branches will be shaken and fallen fruits will be collected from the ground. At the interval of 15-20 days, the collection of fruits will be also done by well shaken and hooks are used to pluck all the fruits with strong stalks. Finally, the late maturity fruits will be collected from natural falling on the ground due to wind and bird/monkey damage.

Tamarind fruits in tribal Patta lands were mostly collected by the owners or leased out and mortgaged to contractors/traders. About 65% of the tree owners collect fruits and process themselves. They sell pulp in weekly markets at Jamunamarathur and Odugathur. They also sell them in markets at Polur, Tiruvannamalai and Vellore at Rs. 50/per kg. About 20% of the trees were given for annual leases to contractors. The tree owners will get Rs. 500 to 1200 per tree depending on tree size and yield. The rates are very low and they never come out of the clutches of the traders. Tamarind pulp in city retail shops sold at Rs. 120 to 140/per kg. The pulp concentrates sold Rs. 170 to 180/kg. Tamarind pulp and seed products have good demand in South India as well as global markets. The Tamarind seed products are being used in other allied industries. In addition to local production the raw Tamarind is being purchased from other states and brought to Tamil Nadu for value addition. Hence, the Tamarind processing industries have a bright future and there is no dearth of raw materials (TASPEF Report, 2012). Generally 4-5 times harvest will be carried out in forest areas, whereas it is 3-4 times in roadside area. More Tamarind processing units, cold storages and mundis are mostly located in Darmapuri, Krishnagiri and Salem districts.

Marketing aims at an effective production system where information is transmitted effectively between production and consumption (Juslin and Lintu, 1997). Marketing needs to be recognized at several stages from forests to final consumers (Lintu, 1986). Marketing shifts from manipulating consumers to genuine consumer's involvement, from telling and selling to communicating and sharing knowledge (McKenna, 1991). The marketing has even greater importance and significance for the society as a whole than for any of the individual beneficiaries of the marketing process. The middlemen are reaping huge profits through the resale of forest produce in markets within and outside the state (Jain, 1989; Lal, 1998). Since, the economic activities are largely concerned with the marketing process, scientific and systematic organization of marketing system is very essential for the healthy development of forest economy of a particular region. According to Blanchard (2010), the most ideal practice of creating demand is to rigorously market the product as a strategy to increase its consumption. In the process of marketing of Tamarind, the difference between the price paid by the consumer and the price received by the producer for an equivalent quantity was defined as "Price Spread." To have a holistic picture of price spread, the data from each marketing system for selected Tamarind are tabulated for comparative analysis as below. The price spreads worked out for the marketing systems I; II and III of Tamarind (Fig. 9.8).

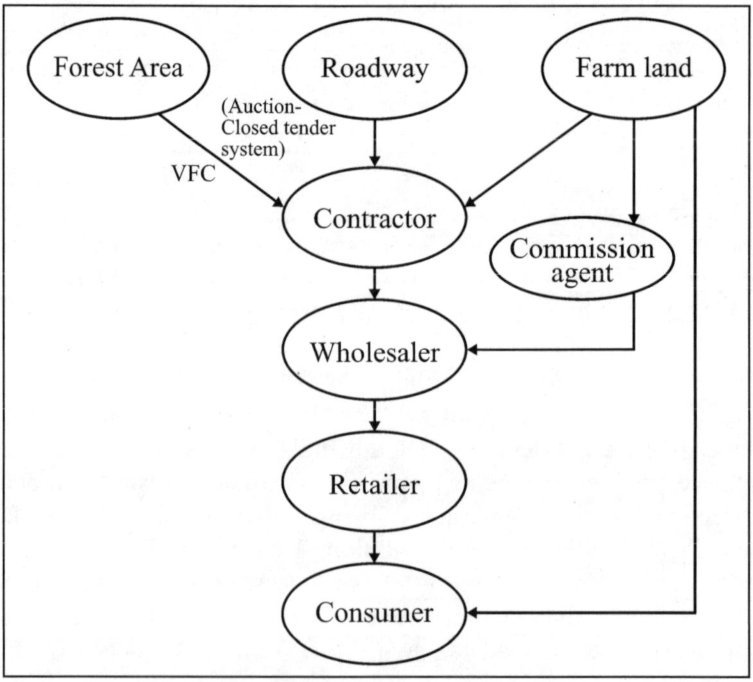

Marketing System I	Forest area - Contractor - Wholesaler - Retailer - Consumer
Marketing System II	Roadway - Contractor - Wholesaler - Retailer - Consumer
Marketing System III	Farm land - Contractor - Wholesaler - Retailer - Consumer
Marketing System IV	Farm land - Commission agent - Wholesaler - Retailer - Consumer
Marketing System V	Farm land - Consumer

Fig 9.8. Marketing System of Tamarind in Tamil Nadu

9.5. PRICE SPREAD ANALYSIS

Information was collected from the individual farmers and traders. The costs include the transport, weighing, loading and unloading, packing, storage, spoilage, commission charges and other expenses incurred for marketing the produce. In the process of marketing of Tamarind, the difference between the price paid by the consumer and the price received by the producer for an equivalent quantity of Tamarind produce was defined as "Price Spread". The profits of the various market functionaries involved in moving the produce from the initial point of production until it reached the ultimate consumer were recorded. In general, Sum-of-Average Gross Margin method was used in the estimation of the price spread. The average gross margins of all the intermediaries were added to obtain the total marketing margin as well as the breakup of the consumer's rupee (Table 9.5).

$$MT = \sum_{i=1}^{n} \frac{[\{S_i - P_i\}]}{Q_i}$$

Where,

MT = Total Marketing Margin

Si = Sale value of a product for i^{th} intermediary

Pi = Purchase value paid by the ith intermediary

Qi = Quantity of the product handled by the ith intermediary

i = 1, 2, 3 … N (Number of intermediaries involved in the value chain)

Table 9.5. Price Spread of Tamarind in the Selected Marketing Systems

Items	MS-I		MS-II		MS-III		MS-IV		MS-V	
	(a)	**(b)**	**(a)**	**(b)**	**(a)**	**(b)**	**(a)**	**(b)**	**(a)**	**(b)**
Forest / road/ farmers	**Forest Area**		**Roadway**		**Farmer**		**Farmer**		**Farmer**	
Net price received by the forest dependent/road/ farmer	1500	18.75			3200	45.71	4000	56.34	6600	94.29
Marketing cost	50	0.63			260	3.71	282	3.97	305	4.36
Profit margin	1318	16.48			340	4.86	68	0.96	95	1.36
Marketing margin	2000	25.00			600	8.57	350	4.93	400	5.71
Contractor										
Price paid by contractor	3500	43.75	3400	43.59	3800	54.29				
Marketing cost	50	0.63	780	10.00	455	6.50				
Profit margin	950	11.88	270	3.46	1245	17.79				
Marketing margin	1000	12.50	1050	13.46	1700	24.29				
Commission Agent (CA)										
Price paid by the CA							4350	61.27		
Marketing cost							225	3.17		
Profit margin							985	13.87		
Marketing margin							1210	17.64		
Wholesaler										
Price paid by the wholesaler	4500	56.25	4450	57.05	5500	78.57	5560	78.31		
Marketing cost	185	2.31	350	4.49	220	3.14	220	3.10		
Profit margin	1815	22.69	1350	17.31	481	6.87	421	5.93		
Marketing margin	2000	25.00	1700	21.79	701	10.01	641	9.03		
Retailer										
Price paid by the retailer	6500	81.25	6150	78.85	6201	88.59	6201	87.33		
Marketing cost	170	2.13	190	2.44	101	1.44	101	1.42		
Profit margin	1330	16.63	1460	18.72	698	9.97	798	11.24		
Marketing margin	1550	18.75	1650	21.15	799	11.41	899	12.67		
Price paid by the consumer	8000	100.00	7800	100.00	7000	100.00	7100	100	7000	100.00
Price spread	6500 (81.25 %)		4400 (56.41 %)		3800 (54.29 %)		3100 (43.66 %)		400 (5.71 %)	

It could be observed that in marketing system I, the net price received by the forest dependent was Rs. 1500/Qtl, which constituted about 18.75 per cent of the consumer price. The marketing cost incurred by the wholesaler was 2.31 per cent of the consumer price. The profit margin was found to be higher for a wholesaler (22.69 per cent of the consumer price). The total marketing margin was 81.25 per cent of the consumers' rupee. The difference between the net price received by the forest dependent and the price paid by the consumer was Rs. 6500/Qtl.

In the marketing system II, the net price received by the contractor was Rs. 3400/Qtl. The cost of marketing was the highest to the contractor, which constituted about 10.00 per cent of the consumer price, followed by wholesaler (4.49 per cent) and retailer (2.44 per cent). The profit margin was found to be higher for the retailer (18.72 per cent), followed by the wholesaler (17.31 per cent) and the contractor (11.88 per cent). The marketing margin was found to be higher for the wholesaler (21.79 per cent), followed by retailer (21.15 per cent). Price spread in system II was Rs. 4400/Qtl, which was better than the price spread of marketing system I.

In marketing system III, the net price received by the farmer was Rs. 3200/Qtl. The cost of marketing was the highest to the contractor, which constituted about 6.50 per cent of the consumer price, followed by the farmer (3.71 per cent), wholesaler (3.14 per cent) and retailer (1.44per cent). The profit margin was found to be higher for the contractor (17.79 per cent), followed by the retailer (9.97 per cent), wholesaler (6.87 per cent) and the farmer (4.86 per cent). The marketing margin was found to be higher for the contractor (24.29 per cent), followed by retailer (11.41 per cent) and wholesaler (10.01 per cent). Price spread in system III was Rs.3800/Qtl, which was better than the price spread of marketing system I and II.

In marketing system IV, the net price received by the farmer was Rs. 4000/Qtl. The cost of marketing was the highest to the farmer, which constituted about 3.97 per cent of the consumer price, followed by the commission agent (3.17 per cent), wholesaler (2.10 per cent) and retailer (1.42 per cent). The profit margin was found to be higher for the commission agent (13.87 per cent), followed by the retailer (11.24 per cent), and wholesaler (5.93 per cent). The marketing margin was found to be higher for the commission agent (17.64 per cent), followed by retailer (12.67per cent) and wholesaler (9.03 per cent). Price spread in system III was Rs.3100/Qtl, which was better than the price spread of marketing system I, II and III.

In marketing system V, the net price received by the farmer was Rs.6600/Qtl and consumer price was Rs 7000/Qtl. The cost of marketing was Rs 305/Qtl (4.36 per cent). Theoretically, the price spread for this system would be zero. However, in practical circumstances, a certain amount of cost is incurred in this system due to transportation cost from the fields and post-harvest losses. This marketing system had a low price spread in (5.71 per cent). The marketing system V was found to be the best system among other marketing systems.

Tamarind has a huge domestic and overseas market and commands a good price from the end-consumers. However, primary producers and collectors get very low prices and the bulk of the value is captured by the middlemen. Krishnagiri, Dharmapuri and Salem are the main markets for Tamarind. Nearly 500- 600 retail marketing shops viz., Mundis and 6-8 cold storage-godowns are located in Chennai, Krishnagiri, Dharmapuri and Salem districts (TASPEF Report, 2011).

Many collectors in Odessa and Madhya Pradesh are forced to sell Tamarind at very low rates to petty traders. Some traders barter Tamarind with low value items resulting in a major loss for the collectors. Many collectors also have credit linkages with the traders and use Tamarind to clear their dues with the highest interest rate. In Andhra Pradesh, the GCC is acting as a monopoly over the Tamarind market in the entire state to focus the development of tribals in the forest area. The GCC is able to procure nearly 6-8 per cent of total Tamarind available in the state. About 20,000 families are getting benefits through bulk sales of Tamarind seed. In Odessa, the Panchayats have empowered to issue transit permits to transport Tamarind, transporters still face problems from the forest check posts. In Madhya Pradesh, the total production of Tamarind in the state is around 1,200 quintals. Siddhi and Chindwara are important Tamarind producing districts. In Odessa, Tamarind is collected mainly from revenue lands and most of the collection is made from the southern districts of the state, such as Gajapati and Rayagada. A large proportion of the collected Tamarind is marketed to Andhra Pradesh. In Karnataka, forest – tribal people – processing units- village fairs-consumer. In some places, farmer- middlemen- processor-retailer-consumer. Bijapur is a major market in Karnataka.

9.6. WHOLESALE PRICE OF TAMARIND

The wholesale price index Tamarind is depicted in Figure 9.8. The base year for index calculation is 2000 which is considered as normal year for Tamarind in Tamil Nadu. The wholesale index in 2013 was 313 which is higher than base year value. This highest index in the recent past years showed that there could be a good price for Tamarind. This will augment the profitability of Tamarind plantation in Tamil Nadu (Fig. 9.9)

There is a persistent rise of Tamarind prices in the last two years due to the reduction of potential area under the tree-crop, reduction of production level, huge export demand and local demand. Some time, the price may go up to Rs. 180 per kg at retail outlets for a processed Tamarind (without seeds). The price is based on supply and demand. In December, 2009, the price of the Tamarind was Rs 27.50 per kg, whereas it was Rs 76.67 per kg in August, 2009. According to the Agriculture Ministry, the modal price or the rate at which most trades took place in Tamarind (with seeds) is currently Rs. 3,400/quintal at Hindupur market in Andhra Pradesh. The fine variety is ruling at Rs. 7,900/quintal. In Bangalore, the modal price for Tamarind is currently Rs. 10,000/quintal in 2014.

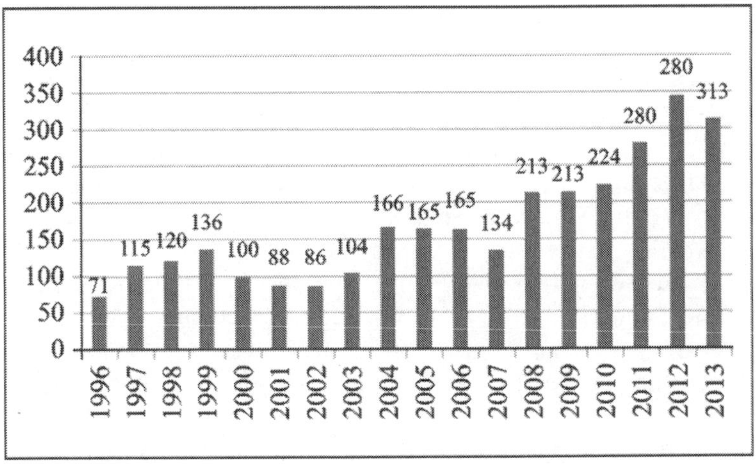

Fig 9.9. Wholesale Price Index of Tamarind in Tamil Nadu

9.7. INTERNATIONAL TRADE OF TAMARIND

The export of the Tamarind was increased from 9384 tonnes to the value of Rs. 6.32 crore in 1988-89 to 17500 tonnes (worth of Rs. 80 crore) in 2010-11 (Fig. 9.10). The price of Tamarind is highly volatile and season oriented fluctuation. Tamarind prices may be high in the end of September when prices and low after Diwali due to a new arrival in December. There are export opportunities in the Far East which can be tapped for Tamarind (Ramesh, 2010). Among foreign markets, Singapore, Malaysia, Germany, United Kingdom and United States of America are the largest importers of Tamarind. If demand from the Gulf, Pakistan, Bangladesh and Malaysia is higher, the wholesale price may increase and the crop failure in Indonesia, Burma and Thailand will also increase the price (Fig. 9.10).

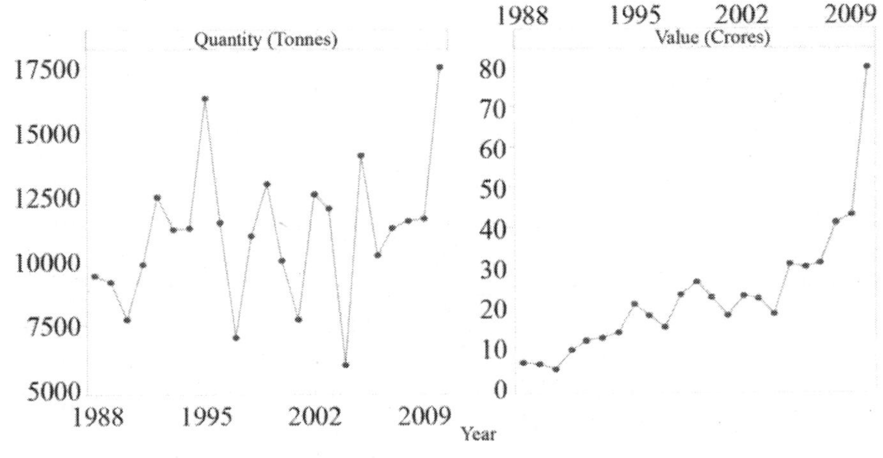

Fig. 9.10. Export Quantity and Value of Tamarind in India

9.8. CONCLUSION AND POLICY OPTIONS

There is a potential scope for Tamarind production and trade promotion in India and Tamil Nadu. The reduction of Tamarind area is due to the developmental activities of urbanization. This can be increased in establishing scientific and profitable plantations in forest, common and private lands. It should be mapped with land use pattern. Still, there are possibilities to boost area in tree outside the forest area. The highest productivity districts should be targeted. The production should be enhanced with high yielding and location specific varieties to be developed and planted on a large scale. Natural phenomenon such as drought, heavy winds and heavy rainfall also affects the flowers and fruit setting process. It should be mitigated with appropriate technological intervention such as spraying of chemicals to improve the drought resistance and fruit settings. It is found that the availability of Tamarind fruits in forest areas is very much higher than the present collections. The scientific harvesting methods are to be given for harvesting persons and collectors. Enough protection should be given in the fruiting season (November – February).

The efficient marketing channel should be informed to producers in order to maximize their share in consumer's price. The industrial tie with processing / value added industries should be established with all VFC and farmers. The contract farming may be also introduced in Tamarind farming. The SHGs in tribal area can be established and trained for value addition of Tamarind. The proper marketing information system should be established to improve the marketing of Tamarind and value added products of Tamarind. There is a potential for export of Tamarind. Hence, the VFC members / farmers/ SHGs/ unemployed youths are to be given trainings for export of Tamarind and trainings should be given to the tribal and farmers on linking of producers with appropriate markets and establish a farmer producer organization (FPO) for Tamarind.

A large volume of fruits and seeds is wasted due to the lack of processing and storage technologies. Available technologies have not been disseminated among growers. Processed pulp is of low quality and does not meet the standards of the international market. Alternative low-cost processing technologies are not readily available. The tremendous industrial potentiality of Tamarind has not been explored adequately. Lack of both local and international market information restricts expansion of crop and product diversification (Varmudy, 2013). This will improve the supply chain and value chain of Tamarind to improve its role in the socio economic development of the society and nation. A strong linkage is needed to have a sustainable management and marketing of Tamarind at state and national levels to facilitate the development of a database of Tamarind and their improved marketing information network for directing the Tamarind availability from conventional forest and outside forest areas to most appropriate and rewarding end-uses.

REFERENCES

Blanchard, D. (2010). Supply chain management best practices. John Wiley & Sons.

FGLG Report (2008). Non-Timber Forest Products and Forest Governance, Synthesis Report by Forest Governance Learning Group-India 1-2.

Hand Book on Horticulture Statistics (2014). Government of India, Ministry of Agriculture, Department of Agriculture and Cooperation, New Delhi

Jain, A.K. (1989). Marketing of forest products in Balaghat district. Forest Economy of India. 273-311.

Joshua N Daniel (2006). Analysis of Economical Characteristics of Value Chains of Three. The International Centre for Underutilized Crops.

Juslin, H and Lintu, L. (1997). Responses to changes in demand and supply of forest products through improved marketing. Proceedings of the XI World Forestry Congress, Vol. 4, 121-135.

Lal, R.B. (1998). Marketing information system for non-wood forest products in India, Marketing Monitoring of

Lintu, L. (1986). Marketing on the forestry sector. *UNASYLVA*. 38 (153), 10-19.

Mc Kenna, R. (1991). Marketing is Everything. Harvard Business Review, Jan-Feb.1991, 65-79

Ramesh Chand (2010). Estimation of Marketing Efficiency of Horticultural Commodities under Different Supply Chains in India. National Centre for Agricultural Economics and Policy Research, New Delhi, 172-173.

Season and Crop Report (2013). Department of Economics and Statistics Chennai, TAMIL NADU

Singh, Dheeraj, Lubsang Wangehu and Surendra Kumar Moond (2007). Processed Products of Tamarind. Natural Product Radiance, (4): 315-321.

TASPEF Report (2011). Vision for the management of non-timber forest products (NTFPs) in Tamil Nadu 2011-12. A research study on NTFP Resources relating to Edible Plant Industry and Tannin Industry, Tamil Nadu Association of Senior Professionals of Environment and Forests (TASPEPs), Chennai, 5-113

Varmudy, Vigneshwara (2013). Tamarind's potential marred by neglect. Market survey, FACTS FOR YOU, January 2013, pp 22-24.

www.indiastat.com

■■■■